The *Molecular World*

Alkenes and Aromatics

edited by

Peter Taylor

This publication forms part of an Open University course, S205 *The Molecular World*. Most of the texts which make up this course are shown opposite. Details of this and other Open University courses can be obtained from the Call Centre, PO Box 724, The Open University, Milton Keynes MK7 6ZS, United Kingdom: tel. +44 (0)1908 653231, e-mail ces-gen@open.ac.uk.

Alternatively, you may visit the Open University website at http://www.open.ac.uk where you can learn more about the wide range of courses and packs offered at all levels by The Open University.

The Open University, Walton Hall, Milton Keynes, MK7 6AA

First published 2002

Edited, designed and typeset by The Open University.

Published by the Royal Society of Chemistry, Thomas Graham House, Science Park, Milton Road, Cambridge CB4 0WF, UK.

Printed in the United Kingdom by Bath Press Colourbooks, Glasgow.

ISBN 0 85404 680 1

A catalogue record for this book is available from the British Library.

1.1

s205book 7 i1.1

The Molecular World

This series provides a broad foundation in chemistry, introducing its fundamental ideas, principles and techniques, and also demonstrating the central role of chemistry in science and the importance of a molecular approach in biology and the Earth sciences. Each title is attractively presented and illustrated in full colour.

The Molecular World aims to develop an integrated approach, with major themes and concepts in organic, inorganic and physical chemistry, set in the context of chemistry as a whole. The examples given illustrate both the application of chemistry in the natural world and its importance in industry. Case studies, written by acknowledged experts in the field, are used to show how chemistry impinges on topics of social and scientific interest, such as polymers, batteries, catalysis, liquid crystals and forensic science. Interactive multimedia CD-ROMs are included throughout, covering a range of topics such as molecular structures, reaction sequences, spectra and molecular modelling. Electronic questions facilitating revision/consolidation are also used.

The series has been devised as the course material for the Open University Course S205 *The Molecular World*. Details of this and other Open University courses can be obtained from the Course Information and Advice Centre, PO Box 724, The Open University, Milton Keynes MK7 6ZS, UK; Tel +44 (0)1908 653231; e-mail: ces-gen@open.ac.uk. Alternatively, the website at www.open.ac.uk gives more information about the wide range of courses and packs offered at all levels by The Open University.

Further information about this series is available at www.rsc.org/molecularworld.

Orders and enquiries should be sent to:

Sales and Customer Care Department, Royal Society of Chemistry, Thomas Graham House, Science Park, Milton Road, Cambridge, CB4 0WF, UK

Tel: +44 (0)1223 432360; Fax: +44 (0)1223 426017; e-mail: sales@rsc.org

The titles in *The Molecular World* series are:

THE THIRD DIMENSION
    edited by Lesley Smart and Michael Gagan

METALS AND CHEMICAL CHANGE
    edited by David Johnson

CHEMICAL KINETICS AND MECHANISM
    edited by Michael Mortimer and Peter Taylor

MOLECULAR MODELLING AND BONDING
    edited by Elaine Moore

ALKENES AND AROMATICS
    edited by Peter Taylor and Michael Gagan

SEPARATION, PURIFICATION AND IDENTIFICATION
    edited by Lesley Smart

ELEMENTS OF THE p BLOCK
    edited by Charles Harding, David Johnson and Rob Janes

MECHANISM AND SYNTHESIS
    edited by Peter Taylor

The Molecular World Course Team

Course Team Chair
Lesley Smart

Open University Authors
Eleanor Crabb (Book 8)
Michael Gagan (Book 3 and Book 7)
Charles Harding (Book 9)
Rob Janes (Book 9)
David Johnson (Book 2, Book 4 and Book 9)
Elaine Moore (Book 6)
Michael Mortimer (Book 5)
Lesley Smart (Book 1, Book 3 and Book 8)
Peter Taylor (Book 5, Book 7 and Book 10)
Judy Thomas (*Study File*)
Ruth Williams (skills, assessment questions)

Other authors whose previous contributions to the earlier courses S246 and S247 have been invaluable in the preparation of this course: Tim Allott, Alan Bassindale, Stuart Bennett, Keith Bolton, John Coyle, John Emsley, Jim Iley, Ray Jones, Joan Mason, Peter Morrod, Jane Nelson, Malcolm Rose, Richard Taylor, Kiki Warr.

Course Manager
Mike Bullivant

Course Team Assistant
Debbie Gingell

Course Editors
Ian Nuttall
Bina Sharma
Dick sharp
Peter Twomey

CD-ROM Production
Andrew Bertie
Greg Black
Matthew Brown
Philip Butcher
Chris Denham
Spencer Harben
Peter Mitton
David Palmer

BBC
Rosalind Bain
Stephen Haggard
Melanie Heath
Darren Wycherley
Tim Martin
Jessica Barrington

Course Reader
Cliff Ludman

Course Assessor
Professor Eddie Abel, University of Exeter

Audio and Audiovisual recording
Kirsten Hintner
Andrew Rix

Design
Steve Best
Vicki Eaves
Carl Gibbard
Sarah Hack
Lee Johnson
Mike Levers
Sian Lewis
John Taylor
Howie Twiner

Library
Judy Thomas

Picture Researchers
Lydia Eaton
Deana Plummer

Technical Assistance
Brandon Cook
Pravin Patel

Consultant Authors
Ronald Dell (*Case Study:* Batteries and Fuel Cells)
Adrian Dobbs (Book 8 and Book 10)
Chris Falshaw (Book 10)
Andrew Galwey (*Case Study:* Acid Rain)
Guy Grant (*Case Study:* Molecular Modelling)
Alan Heaton (*Case Study:* Industrial Organic Chemistry, *Case Study:* Industrial Inorganic Chemistry)
Bob Hill (*Case Study:* Polymers and Gels)
Roger Hill (Book 10)
Anya Hunt (*Case Study:* Forensic Science)
Corrie Imrie (*Case Study:* Liquid Crystals)
Clive McKee (Book 5)
Bob Murray (*Study File*, Book 11)
Andrew Platt (*Case Study:* Forensic Science)
Ray Wallace (*Study File*, Book 11)
Craig Williams (*Case Study:* Zeolites)

CONTENTS

PART 1 ADDITION – PATHWAYS AND PRODUCTS

1   ELECTROPHILIC ADDITION REACTIONS OF ALKENES   11

1.1   Introduction   11

1.2   Addition of HX   13

1.3   Addition of halogens and related compounds   18

1.4   Summary of Section 1   25

2   OTHER USEFUL ADDITION REACTIONS   28

2.1   *syn*-Additions; hydrogenation   28

2.2   Summary of Section 2   29

LEARNING OUTCOMES   32

QUESTIONS: ANSWERS AND COMMENTS   33

FURTHER READING   36

ACKNOWLEDGEMENTS   36

PART 2 AROMATIC COMPOUNDS

1	INTRODUCTION	39
2	THE STRUCTURE AND STABILITY OF BENZENE	42
	2.1 Summary of Sections 1 and 2	44
3	ELECTROPHILIC AROMATIC SUBSTITUTION REACTIONS	45
	3.1 General principles	45
	3.1.1 Summary of Section 3.1	47
	3.2 Nitration	48
	3.3 Halogenation	50
	3.4 Sulfonation	51
	3.5 Friedel–Crafts reactions	53
	3.5.1 Friedel–Crafts alkylation	54
	3.5.2 Friedel–Crafts acylation	58
	3.6 Summary of Sections 3.2–3.5	59
4	THE EFFECTS OF SUBSTITUENTS	61
	4.1 Summary of Section 4	67
5	DIAZONIUM SALTS	68
	5.1 From coal tar to dyes	68
	5.2 Coupling reactions of diazonium salts	72
	5.3 Substitution reactions of diazonium salts	73
	5.4 Summary of Section 5	74
	APPENDIX SUMMARY OF REACTIONS USEFUL IN THE SYNTHESIS OF AROMATIC COMPOUNDS	75
	LEARNING OUTCOMES	77
	QUESTIONS: ANSWERS AND COMMENTS	78
	FURTHER READING	88
	ACKNOWLEDGEMENTS	88

PART 3 A FIRST LOOK AT SYNTHESIS

1   STRATEGY FOR THE DISCOVERY OF NEW DRUGS   91

   1.1   Neurotransmitters and receptors   93

2   COMPOUNDS THAT MIMIC THE ACTION OF
   NORADRENALINE — AGONISTS   96

   2.1   Summary of Sections 1 and 2   100

3   THE TARGET: β-AMINOALCOHOLS   101

4   PLANNING THE SYNTHESIS OF PSEUDOEPHEDRINE 102

   4.1   General issues   102

   4.2   Some golden rules   103

   4.3   The next stage   105

   4.4   Where will it all end?   108

   4.5   Summary of Sections 3 and 4   110

5   CARRYING OUT THE SYNTHESIS   111

   5.1   Preparation of Z-1-phenyl-1-propene   111

   5.2   How much do we have?   113

   5.3   Preparation of the bromoalcohol   114

   5.4   Making the oxirane and pseudoephedrine   117

   5.5   How efficient was the synthesis?   120

   5.6   Summary of Section 5   121

6   THE SYNTHESIS OF β-AMINOALCOHOLS   122

7   A DIFFERENT WAY OF LOOKING AT SYNTHETIC
   EFFICIENCY   124

   7.1   'Green chemistry'   124

   7.2   'Green chemistry' in action   126

   7.3   Summary of Sections 6 and 7   129

LEARNING OUTCOMES   130

QUESTIONS: ANSWERS AND COMMENTS   131

FURTHER READING   136

ACKNOWLEDGEMENTS   136

CASE STUDY: INDUSTRIAL ORGANIC CHEMISTRY

1	**INTRODUCTION**	**139**
1.1	The chemical industry	139
1.2	Large- and small-scale production	143
1.3	Sub-division of the organic chemicals industry	143
2	**PETROCHEMICALS**	**144**
2.1	Theoretical considerations	144
2.2	Characteristics of the petrochemical sector	146
2.3	Energy considerations	147
2.4	Chemical considerations	147
	2.4.1 Cracking	150
	2.4.2 Reforming	151
	2.4.3 Building blocks	151
	2.4.4 Ethylene (ethene)	152
	2.4.5 Propylene (propene)	152
	2.4.6 Buta-1,3-diene	152
	2.4.7 Benzene	156
	2.4.8 Toluene (methylbenzene)	157
	2.4.9 Xylenes (dimethylbenzenes)	159
2.5	Environmental concerns	160
2.6	Location of plants	161
3	**SPECIALITY AND FINE CHEMICALS**	**162**
3.1	Some typical fine chemicals	162
	3.1.1 Fusilade™	162
	3.1.2 Penicillins	164
	3.1.3 Procion dyes	165
	3.1.4 Kevlar	166
3.2	Characteristics of the speciality and fine chemicals sector	167
3.3	Chemical considerations	169
3.4	Location of plants	169
4	**WHAT HAPPENS WHEN THE FOSSIL FUELS RUN OUT?**	**170**
	FURTHER READING	**172**
	ACKNOWLEDGEMENTS	**172**
	INDEX	**173**

Part 1

Addition –
pathways and products

edited by Peter Taylor

based on
Elimination and Addition: pathways and products
by Richard Taylor and Peter Taylor

ELECTROPHILIC ADDITION REACTIONS OF ALKENES

1

1.1 Introduction

Addition reactions are those in which atoms or groups add to a molecule containing a double or triple bond, thereby reducing the degree of unsaturation; they are the reverse of elimination reactions. Some typical examples of addition reactions are shown below:

$$H_2C=CH_2 + Br_2 \longrightarrow BrCH_2CH_2Br \qquad (1.1)$$

$$CH_3C\equiv CH + 2HBr \longrightarrow CH_3CBr_2CH_3 \qquad (1.2)$$

$$\underset{H_3C}{\overset{H_3C}{>}}C=O + HCN \longrightarrow H_3C-\underset{CH_3}{\overset{OH}{\underset{|}{\overset{|}{C}}}}-CN \qquad (1.3)$$

$$CH_3C\equiv N + 2H_2 \xrightarrow{\text{catalyst}} CH_3CH_2NH_2 \qquad (1.4)$$

We shall concentrate on addition to alkenes. Although alkynes tend to undergo the same types of reaction, we shall not discuss their reactions in detail.

⬤ Alkenes generally react by ionic mechanisms involving *nucleophiles* and *electrophiles*. Give definitions for each of these terms.

⬤ A *nucleophile* can be defined as a species possessing at least one non-bonded pair of electrons, which ultimately forms a new bond to carbon. An *electrophile* is a positively charged or positively polarized species that reacts with a nucleophile.

Alkenes generally provide the nucleophilic component of the addition. You may find it hard to picture how an alkene can act as a nucleophile. Figure 1.1 shows the bonding picture of a carbon–carbon double bond. Carbon–carbon double bonds are made up of a strong σ bond plus a weaker π bond. The two electrons in the π bond dominate the chemistry of alkenes. They can be thought of as providing a negatively charged cloud of electrons above and below the plane of the carbon atom framework. This electron-rich centre repels nucleophiles and attracts electrophiles.

Figure 1.1
The electron distribution in ethene after overlap of the two p orbitals not used in forming σ molecular orbitals.

---

* This symbol, 🖳, indicates that this Figure is available in WebLab ViewerLite™ on the CD-ROM associated with this Book.

BOX 1.1 The use of curly arrows

This is a good place for a brief reprise of curly arrow notation. Make sure you always use arrows with precision, as here:

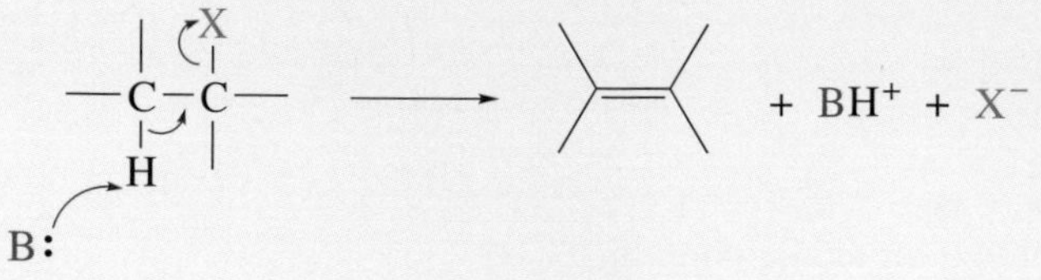

Place each curly arrow carefully to depict movement of an electron *pair*, with the tail at its origin and the head at its destination. Remember that origin and destination can each have only one of two locations: *between* two atoms (bonded pair) or *on* one atom (an unshared or non-bonding pair). You should know that bonding pairs are depicted as lines in organic structures and that non-bonding pairs on atoms such as nitrogen, oxygen and halogens are usually not shown at all. However, if you are unsure about the non-bonded pairs in reaction mechanisms it is a good idea to draw the non-bonding electrons in as dots, so you can keep track of the electrons as they move during a reaction.

So it is the pair of electrons in the π bond that acts as the nucleophile in the reactions of alkenes. Alkenes are certainly electron-rich, but they do not contain a non-bonded pair of electrons. However, although the π electrons are bonding electrons, they do react with electrophiles, as you will see. This is because the π electrons are polarizable; that is, they are far enough from the carbon nuclei to be susceptible to the influence of electrophiles.

One of the most characteristic reactions of alkenes is electrophilic addition, as exemplified by the addition of halogens (X_2) and hydrogen halides (HX) across the double bond:

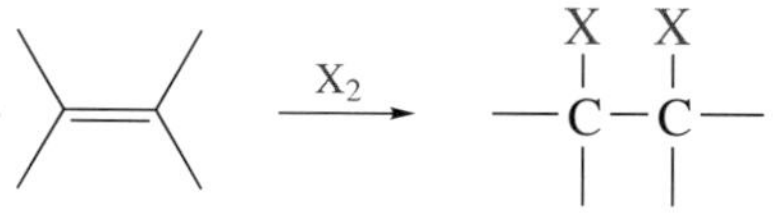

$$X = Cl, \text{ Br or I} \quad (1.5)$$

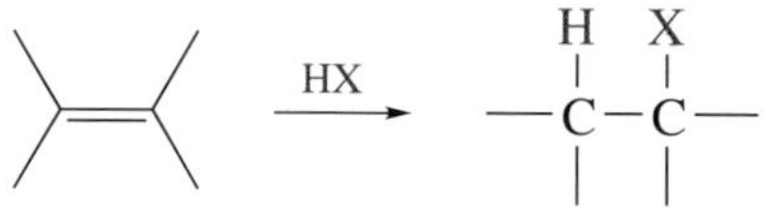

$$(1.6)$$

These reactions can be shown to proceed by a two-step mechanism, in which the first step involves reaction between the alkene and an electrophile. Reaction 1.7 shows the simplest form of this mechanism that is encountered. Notice that although this reaction is called an electrophilic addition reaction, the alkene is a nucleophile. This is because reactions are generally named after the nature of the reagent, and in this case the reagent is electrophilic.

$$(1.7)$$

Look at the mechanism of the electrophilic addition reaction carefully, and try to understand the changes in the bonding.

Two of the electrons from the π system of the alkene form a new bond to the electrophile, which is given the symbol E^+. The carbocation intermediate formed in this first step then reacts with a nucleophile, Nu^-, to give the reaction product. So, in order for reactions such as this to occur, an alkene must be treated with a reagent that provides both an electrophile and a nucleophile.

1.2 Addition of HX

Reactions 1.8, 1.9 and 1.10 show that hydrogen iodide, hydrogen bromide and hydrogen chloride, respectively, all add to alkenes:

$$H_2C{=}CH_2 + HI \longrightarrow CH_3CH_2I \qquad (1.8)$$

$$CH_3CH{=}CHCH_3 + HBr \longrightarrow CH_3CH_2CHBrCH_3 \qquad (1.9)$$

$$(CH_3)_2C{=}CH_2 + HCl \longrightarrow (CH_3)_3CCl \qquad (1.10)$$

● Think back to the mechanism that we proposed for electrophilic addition. Which species do you think acts as the electrophile in Reactions 1.8, 1.9 and 1.10?

● All the hydrogen halides ionize as H^+ and X^-. The proton, H^+, is a strong electrophile.

So the mechanisms of Reactions 1.8 and 1.9 are straightforward:

$$(1.11)$$

$$(1.12)$$

In principle, the addition of an electrophile to an alkene can lead to two different carbocation intermediates:

$$(1.13)$$

However, in Reactions 1.8 and 1.9 the two alkenes are symmetrically substituted, so the same carbocation is produced no matter which carbon–hydrogen bond is formed in the first step. For example:

$$H_2C=CH_2 \xrightarrow{H^+} H_2\overset{+}{C}-CH_3 \tag{1.14}$$

or

$$H_2C=CH_2 \xrightarrow{H^+} H_3C-\overset{+}{C}H_2 \tag{1.15}$$

● In principle, how many carbocation intermediates can be formed from the protonation of 2-methylpropene, $(CH_3)_2C=CH_2$?

● This is an unsymmetrical alkene, so two distinct carbocation intermediates are possible, depending on whether the proton bonds to the central atom of the alkene or to the terminal carbon atom of the double bond. In theory, therefore, this reaction could lead to *two* products, **1.1** and **1.2**:

$$
\begin{array}{ccc}
\underset{H_3C}{\overset{H_3C}{>}}C=CH_2 & \xrightarrow{H^+} & \underset{H_3C}{\overset{H_3C}{>}}\overset{+}{C}-CH_3 \quad \text{or} \quad H_3C-\underset{CH_3}{\overset{H}{\underset{|}{\overset{|}{C}}}}-\overset{+}{C}H_2
\end{array}
$$

$$\downarrow Cl^- \qquad\qquad\qquad \downarrow Cl^-$$

$$(CH_3)_3CCl \qquad\qquad (CH_3)_2CHCH_2Cl$$

1.1 💻 **1.2** 💻   SCHEME 1.1

In practice, the only product is 2-chloro-2-methylpropane (**1.1**). The reason for this predominance is apparent when the relative stabilities of the two intermediate carbocations are considered.

● Which is more stable, a tertiary or a primary carbocation?

● The order of carbocation stabilities is

   tertiary > secondary > primary > methyl (see Box 1.2)

A tertiary carbocation is more stable than a primary one because of the inductive donating effect of the three alkyl groups attached to the charged carbon atom.

BOX 1.2 Stability

There is a slight ambiguity in the way we use the word stable here. In the context of this Book we use it to refer to thermodynamic stability. So when we say that a tertiary carbocation is more stable than a primary carbocation we mean that, relative to the starting material, they have a lower $\Delta G_f^{\ominus}$, and are thus formed more readily.

The other use of the word stable is to imply that something is unreactive, a meaning that certainly does not apply to carbocations!

In electrophilic addition reactions the major product arises from the more-stable carbocation intermediate, because that reaction pathway has the lower energy of activation (see Figure 1.2). So the reactions are kinetically controlled; that is, the major product is the one that is formed faster. Other unsymmetrically substituted alkenes also give the major product by way of the more-stable carbocation intermediate. Thus, Reaction 1.16 proceeds via a secondary, rather than a primary, carbocation; and Reactions 1.17 and 1.18 proceeds via a tertiary, rather than a secondary, carbocation.

$$CH_3CH_2CH{=}CH_2 + HI \longrightarrow CH_3CH_2CHICH_3 \tag{1.16}$$

$$(CH_3)_2C{=}CHCH_3 + HBr \longrightarrow (CH_3)_2CBrCH_2CH_3 \tag{1.17}$$

$$(CH_3)_2C{=}CH_2 + HCl \longrightarrow (CH_3)_2CClCH_3 \tag{1.18}$$

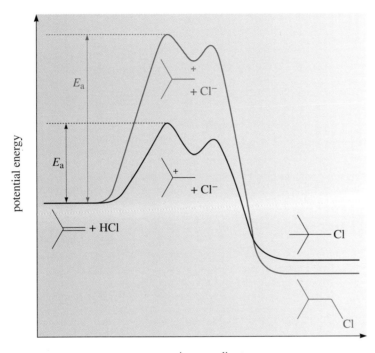

Figure 1.2 Reaction-coordinate diagram (energy profile) for the alternative addition reactions of an unsymmetrical alkene. Note that the favoured black route leads to the less-stable product even though its activation energy, E_a, is less than that for the brown route.

As you can see from Reactions 1.16, 1.17 and 1.18, in order to generate the more-stable carbocation, the proton adds to the less-substituted carbon atom of the double bond — that is, to the one that already has more hydrogen atoms attached.

The generalization that, *in the addition of HX to an unsymmetrical alkene, the hydrogen always adds to the less-substituted carbon atom of the double bond*, was made long before a mechanistic explanation was available. A Russian chemist, Vladimir Markovnikov (Box 1.3), put forward this empirical rule — now known as **Markovnikov's rule** — after studying the products of a number of different HX addition reactions. Markovnikov's rule is sometimes summarized as 'to the one who has, will more be given', because the hydrogen atom from HX goes to the alkene carbon atom with the greater number of hydrogens already attached.

BOX 1.3 Vladimir Vasilevich Markovnikov

Vladimir Vasilevich Markovnikov was born on 22 December 1838 in Nizhny Novgorod, Russia. He, like A. M. Saytzev, studied with A. M. Butlerov in Kazan. Having graduated in 1860, he moved to Germany to work with F. G. C. E. Erlenmeyer and C. Kolbe. He returned to Kazan to take over from Butlerov as Professor of Chemistry. He then moved on to Odessa (1871) and then Moscow (1875), where he died in February 1904. He published his famous rule in 1869, but, since he refused to publish in a foreign language, his work was unknown outside Russia until 1899.

His other notable work included the discovery that branching in alkanes leads to higher melting temperatures, the observation that a hydrogen atom next to the carboxylate group is the most acidic alkyl group of carboxylic acids, and noting that the ease of substitution in radical halogenation is

primary > secondary > tertiary

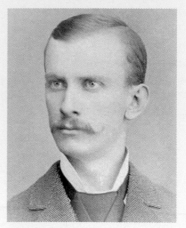

Figure 1.3 Vladimir Vasilevich Markovnikov (1838–1904).

So far, the alkenes that we have considered have been unsymmetrically substituted (for example, $CH_3CH{=}CH_2$) or symmetrically disubstituted (for example, $CH_3CH{=}CHCH_3$). A third possible category consists of alkenes, such as pent-2-ene, in which both carbon atoms of the double bond have the same number of alkyl substituents, but the substituents are different.

- How many products would you expect from the reaction of pent-2-ene, $CH_3CH{=}CHCH_2CH_3$, with HBr?
- You might expect two products, because both possible carbocations formed by the addition of a proton are secondary.

In reactions such as this, a mixture of both possible products usually does result:

$$CH_3CH{=}CHCH_2CH_3 + HBr \longrightarrow CH_3CH_2CHBrCH_2CH_3 + CH_3CHBrCH_2CH_2CH_3 \quad (1.19)$$

Note that Markovnikov's rule applies only to the addition of HX. The earlier mechanistic discussion, however, applies to any electrophilic addition reaction: the major product will always result from the more-stable carbocation. We shall call this the **mechanistic Markovnikov rule** (see also Section 1.3).

- So, what will be the predominant product when HBr is added to $C_6H_5CH{=}CHCH_3$?

$$H^+ + C_6H_5CH{=}CHCH_3 \longrightarrow C_6H_5CH_2\overset{+}{C}HCH_3\ (\mathbf{1.3}) \xrightarrow{Br^-} C_6H_5CH_2CHBrCH_3$$

$$H^+ + C_6H_5CH{=}CHCH_3 \longrightarrow C_6H_5\overset{+}{C}HCH_2CH_3\ (\mathbf{1.4}) \xrightarrow{Br^-} C_6H_5CHBrCH_2CH_3\ (\mathbf{1.5})$$

SCHEME 1.2

In both cases, a secondary carbocation is formed; however, **1.4** will be more stable than **1.3** because in **1.4** the positive charge is adjacent to a phenyl group. This means that the charge can be spread further by *resonance*:

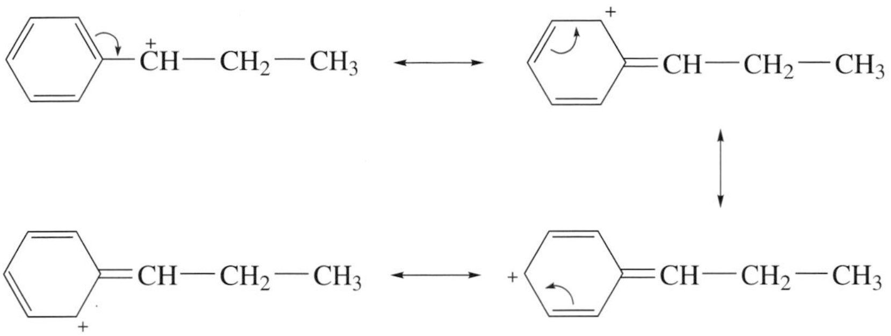

SCHEME 1.3

Since the spreading of charge leads to a more-stable carbocation, **1.5** will be the predominant product.

The hydration of alkenes falls into the same category as the addition of HX. Although this reaction is rarely carried out in the laboratory, it is an important industrial process for preparing alcohols from alkene feedstocks, which in turn are obtained from crude oil*. The alkene is usually passed into a 1 : 1 mixture of sulfuric acid and water. With 2-methylpropene, for example, 60–65% aqueous sulfuric acid is used to prepare 2-methylpropan-2-ol, presumably by way of the corresponding tertiary carbocation:

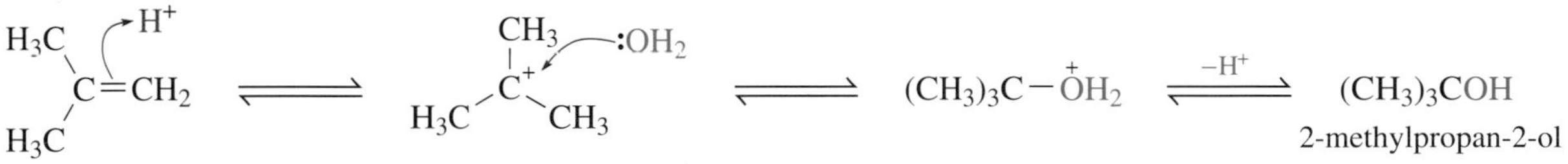

$$(CH_3)_3C-\overset{+}{O}H_2 \xrightarrow{-H^+} (CH_3)_3COH \tag{1.20}$$
2-methylpropan-2-ol

Hydrations are initiated by protons, so Markovnikov's rule is again followed, as you can see from this reaction. The hydration reaction is the reverse of the acid-catalysed elimination of alcohols†: low temperatures and a reaction in aqueous solution favour alcohol formation, whereas elimination is favoured by high temperatures and distillation of the alkene as it is formed.

This balance between addition and elimination can be exploited in order to bring about the migration of a double bond within a molecule. If an alkene is treated with a dilute acid, protonation followed by deprotonation can occur:

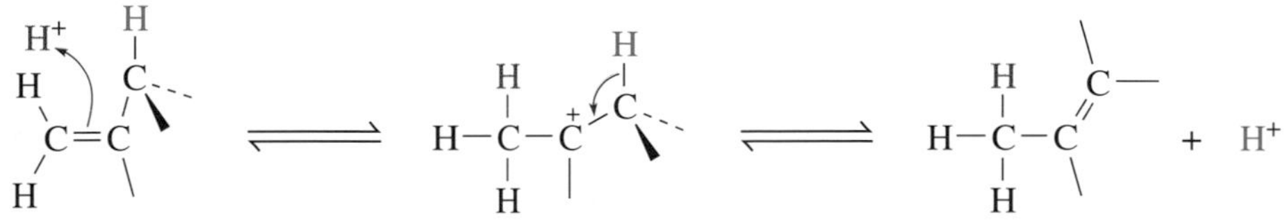

$$\tag{1.21}$$

An equilibrium is established, which will favour the alkene that is more thermodynamically stable. This is usually the most-substituted alkene, so

---

* There is further discussion of the importance of the petrochemicals industry for the synthesis of organic compounds in the Case Study *Industrial Organic Chemistry* at the end of this Book.

† The elimination of water from alcohols is discussed in *Chemical Kinetics and Mechanism*. See Further Reading (p. 36).

this method is sometimes used to convert a less-substituted alkene into a more-substituted one; hence, for example, in Reaction 1.22

$$\text{(Reaction 1.22)} \qquad (1.22)$$

1.3 Addition of halogens and related compounds

Alkenes react readily with chlorine and bromine to produce 1,2-dihalides:

$$\text{(Reaction 1.23)} \qquad X = Cl \text{ or } Br \qquad (1.23)$$

Let's concentrate on the addition of bromine, although all the subsequent discussion applies equally well to the addition of chlorine. The mechanism of this reaction is not immediately obvious because bromine is a non-polar molecule. However, the addition does proceed by an ionic mechanism in which the halogen molecule provides the electrophile. In isolation, bromine is a covalent molecule with a symmetrical electron distribution. However, in the presence of the high electron density of the alkene double bond, polarization of bromine occurs, and one bromine atom becomes electrophilic (Figure 1.4).

By analogy with the addition of HX, the mechanism for the addition of bromine to an alkene can be written as shown in Reaction 1.24:

$$\text{(Reaction 1.24)} \qquad (1.24)$$

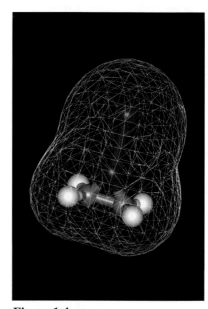

Figure 1.4
Polarization of the Br—Br bond in the presence of a carbon–carbon double bond. The red areas of the surface indicate regions of high electron density, and the blue areas indicate low electron density.

(The analytical applications of this type of reaction are discussed in Box 1.4.)

However, this is not the complete story. In the reaction of *trans*-but-2-ene with bromine, for example, the only product is the diastereomer that results from one bromine atom approaching from one face of the substrate and the other bromine atom approaching from the opposite face:

$$\text{(Reaction 1.25)} \qquad (1.25)$$

BOX 1.4 The alkene–halogen reaction as an analytical tool

Fats and oils are triesters made from glycerol and carboxylic acids — hence their name 'triglycerides':

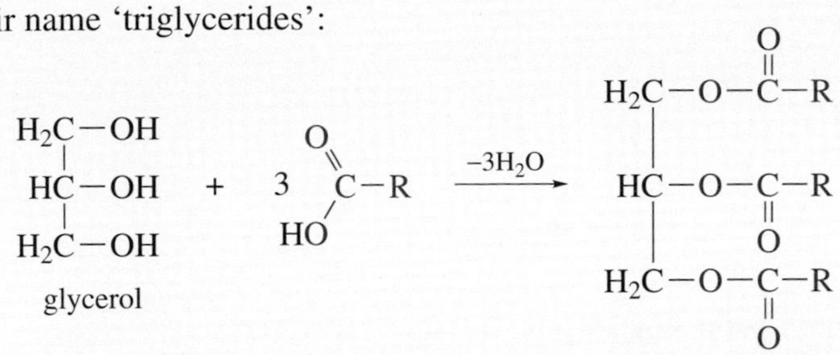

$$
\begin{array}{c}
H_2C-OH \\
HC-OH \\
H_2C-OH \\
\text{glycerol}
\end{array}
\quad + \quad
3 \;
\begin{array}{c}
O \\
\parallel \\
C-R \\
HO
\end{array}
\quad \xrightarrow{-3H_2O} \quad
\begin{array}{c}
O \\
\parallel \\
H_2C-O-C-R \\
| \\
HC-O-C-R \\
\parallel \\
O \\
| \\
H_2C-O-C-R \\
\parallel \\
O
\end{array}
\qquad (1.26)
$$

Nature uses a range of carboxylic acids to make fats and oils, usually long, linear-chain carboxylic acids with an even number of carbon atoms. Saturated fatty acids contain no carbon–carbon double bonds, monounsaturated fatty acids contain one *cis* double bond, and polyunsaturated fatty acids contain more than one *cis* double bond. Hence, linoleic acid (Z,Z-octadeca-9,12-dienoic acid, **1.6**) is a polyunsaturated fatty acid:

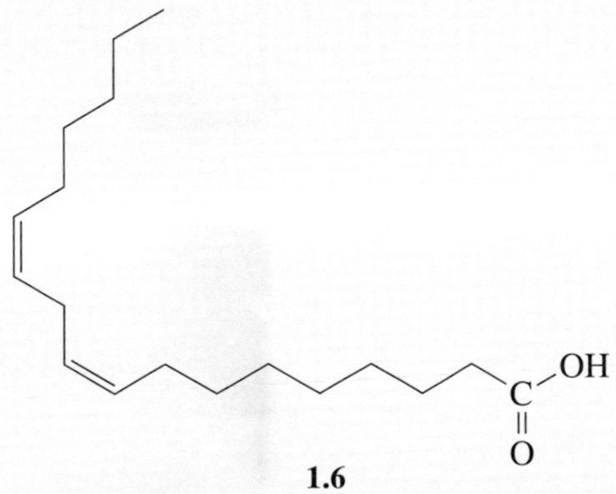

1.6

Saturated fats and oils are thought to be less healthy than monounsaturated fats, which in turn are less healthy than polyunsaturated fats. Thus, it is important to know the amount of unsaturation in a fat or oil. This can be estimated using the 'iodine number'. As double bonds readily react with halogens by electrophilic addition, one way of finding out the amount of unsaturation is to determine the quantity of a halogen that reacts with a particular amount of fat or oil — the greater the amount of halogen, the greater the degree of unsaturation. Since early tests involved the addition of iodine, this has become known as the *iodine number* — the mass, in grams, of iodine that reacts with 100 g of fat or oil. Some typical values are shown in Table 1.1.

Table 1.1
Iodine numbers of some typical fats and oils

Fat or oil	Iodine number
beef fat	35–42
cocoa butter	31–42
olive oil	80–88
corn oil	110–133
sardine oil	185

● Would this result be expected if a simple carbocation intermediate were involved?

By analogy with the S_N1 mechanism, you would expect the bromide ion to attack the carbocation indiscriminately. If this were the case, two products (diastereomers) would result:

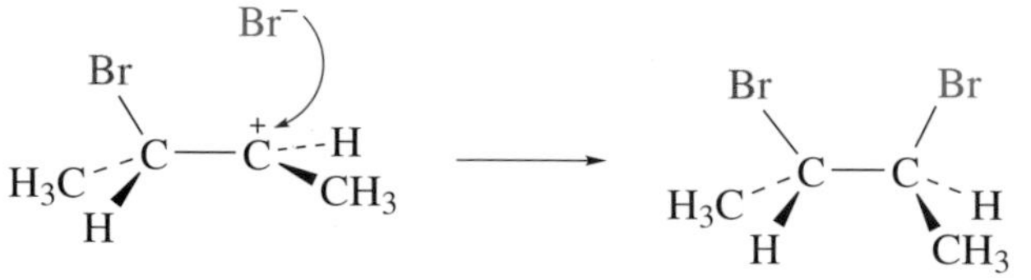

(1.27)

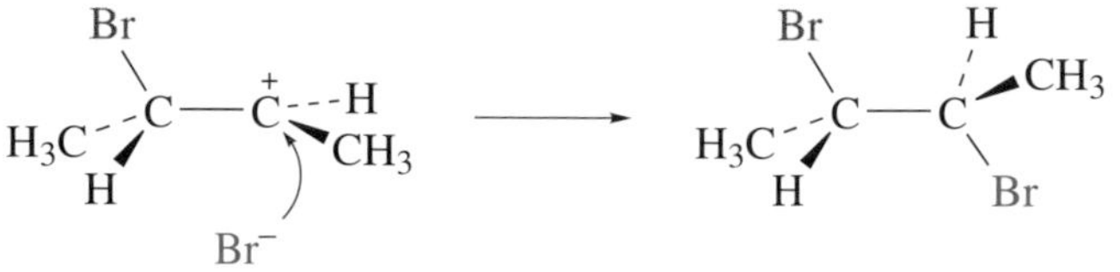

(1.28)

To explain these stereochemical observations, the postulated mechanism has to be slightly modified. One possibility is that, although a carbocation intermediate is still formed, a non-bonded electron pair on the bromine atom is donated to the carbocation to form a three-membered ring containing a positively charged bromine. This is called a **cyclic bromonium ion**. (In general, halogen additions proceed via a cyclic *halonium* ion; thus, if the electrophile were Cl_2, the intermediate would be a *chloronium* ion.)

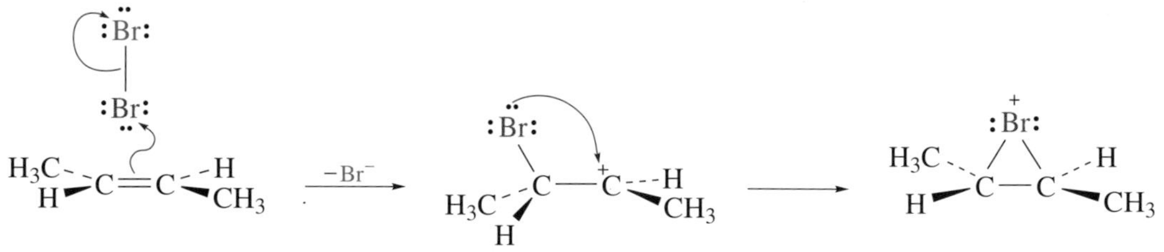

(1.29)

Reaction 1.29 shows the first two steps of the mechanism. The relative disposition of the substituents in the three-membered ring reflects that on the alkene, because the second step in Reaction 1.29 occurs very quickly, before rotation about the central carbon–carbon bond can occur. In fact, it occurs so fast that we usually represent Reaction 1.29 as a single step:

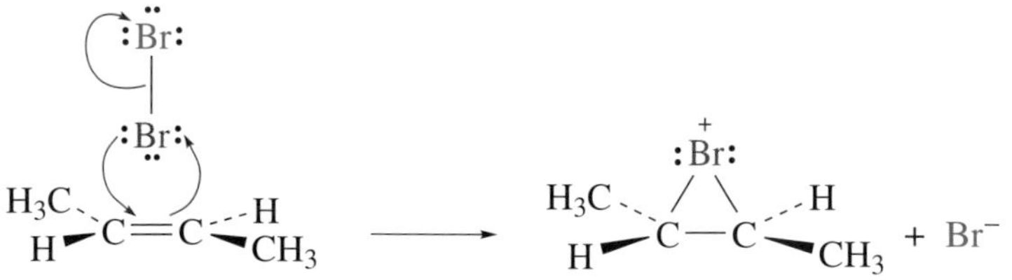

(1.30)

Once the bromonium ion intermediate has been formed, the bromide ion then attacks, opening up the three-membered ring:

(1.31)

This is very much like an S_N2 reaction, and attack occurs from the side opposite the 'leaving group'. As you can see, the overall result is the *anti*-addition of bromine, and the product is *meso*-2,3-dibromobutane.

● Would attack of bromide ion at the other carbon atom of the alkene double bond (as shown in Structure **1.7**) lead to a different product?

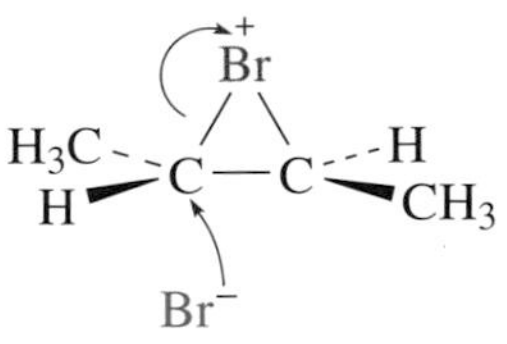

● This again leads to the overall *anti*-addition of bromine, but because of the internal symmetry of the molecule, the same product results, as shown in Scheme 1.4. (The two structures shown are simply two conformations of the same molecule, *meso*-2,3-dibromobutane.)

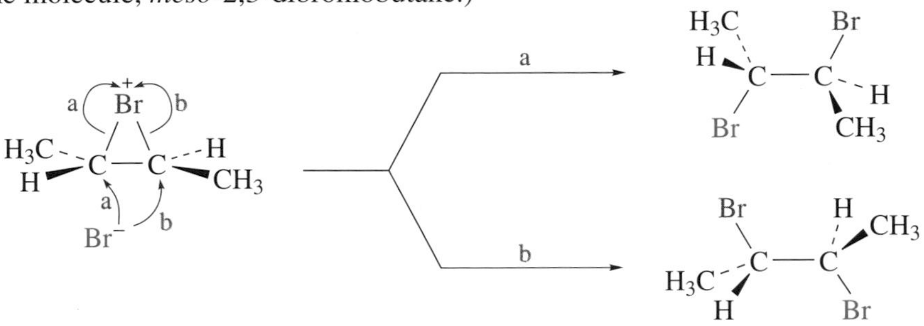

1.7

SCHEME 1.4

STUDY NOTE

If you have access to a molecular model kit like the Orbit system, your understanding of these conformations would be aided by making models. Alternatively, you may view the WebLab ViewerLite representations of these structures on the CD-ROM.

● What would be the final product of addition if, as is equally likely, the initial attack of the bromine molecule was from *below* the plane of the alkene (as in **1.8** rather than from *above* as in Reaction 1.30)?

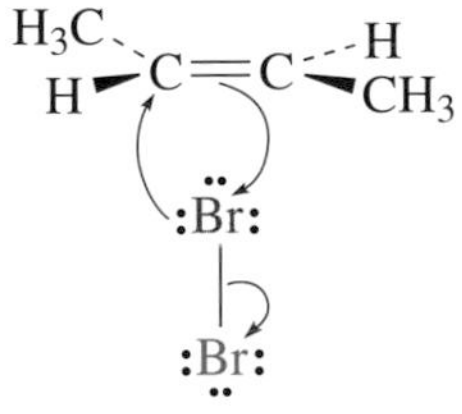

1.8

● This time, the bromide ion subsequently attacks the bromonium ion so formed at either carbon atom from *above,* to give the same *meso* product, as shown in Scheme 1.5:

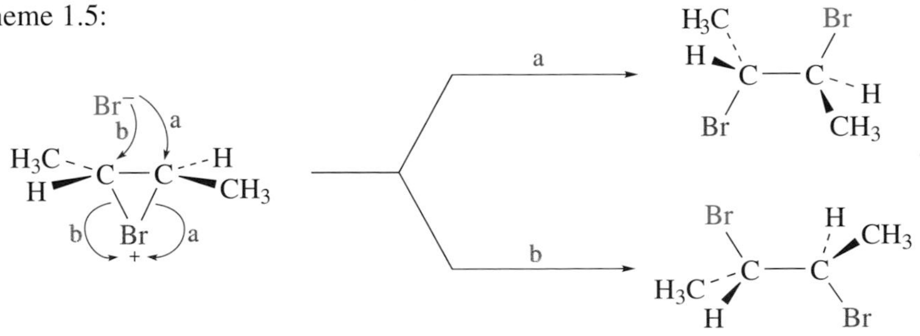

SCHEME 1.5

So, symmetrically substituted *trans* alkenes lead only to *meso* products. However, symmetrically substituted *cis* alkenes react with bromine to produce enantiomeric products. This can be seen from the reaction of *cis*-but-2-ene:

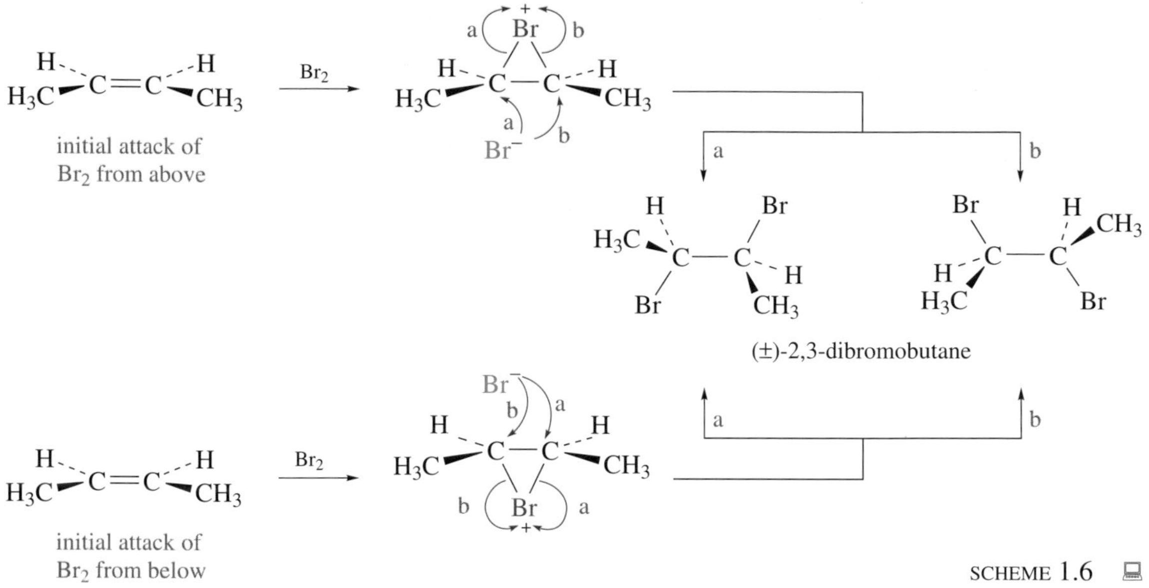

SCHEME 1.6

The notion of an intermediate bromonium ion to explain the observed *anti*-addition of bromine to alkenes was first suggested in 1938. As yet, bromonium ions have not been isolated, although they have been detected using spectroscopy. So, the two-step mechanism for electrophilic addition is substantiated by stereochemical and spectroscopic evidence.

Other evidence, which is also incompatible with a one-step addition mechanism, is obtained if the bromination reaction is carried out in the presence of other nucleophiles. Bromination reactions are normally carried out in an inert solvent such as tetrachloromethane. However, use of a solvent that is itself nucleophilic, such as methanol, results in a mixture of products:

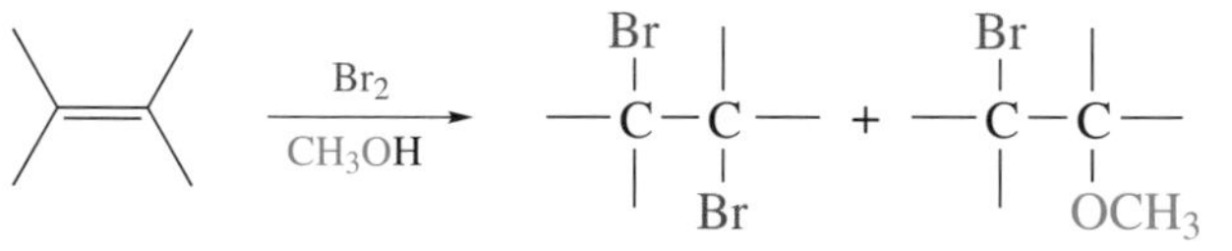

$$(1.32)$$

● How would you account for the formation of the bromomethoxy compound in this reaction?

● Methanol competes with bromide ion as the nucleophile to capture the bromonium ion intermediate, and yields an ether as a byproduct:

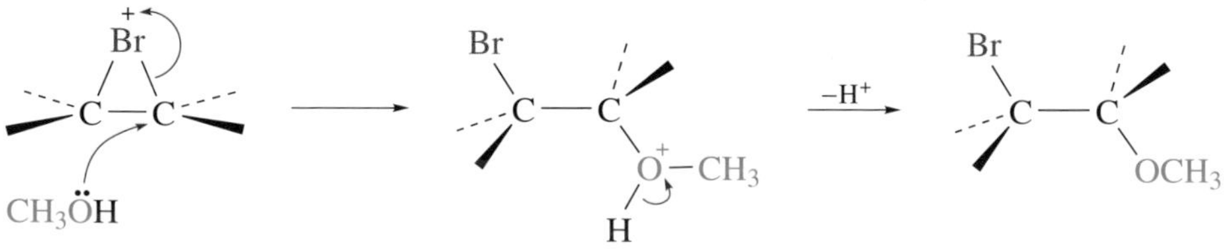

$$(1.33)$$

This type of reaction is often used to prepare bromoethers. To ensure that a good yield of the bromoether is obtained, rather than the dibromo product, the reaction is carried out in dilute solution. This means that methanol is present in far greater concentration than bromide ion. A similar reaction results if water is the solvent:

$$H_2C{=}CH_2 \quad \xrightarrow[H_2O]{Br_2} \quad BrCH_2CH_2OH \;+\; BrCH_2CH_2Br \qquad\qquad (1.34) \quad \square$$

$$\underset{\substack{\text{major product,}\\ \text{a bromohydrin}}}{BrCH_2CH_2OH} \qquad \underset{\text{minor product}}{BrCH_2CH_2Br}$$

This reaction gives rise to bromoalcohols, or **bromohydrins,** as they are sometimes called, you will meet these compounds again in Part 3, where we discuss their importance in the synthesis of pseudoephedrine.

● How many bromohydrins can result from Reaction 1.35?

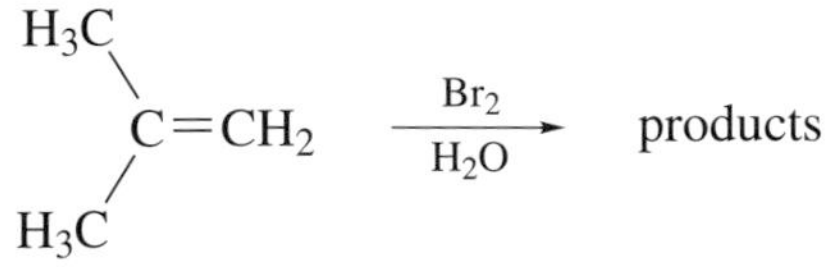

$$\underset{H_3C}{\overset{H_3C}{>}}C{=}CH_2 \quad \xrightarrow[H_2O]{Br_2} \quad \text{products} \qquad\qquad (1.35)$$

○ In theory, two: compounds **1.9** and **1.10**,

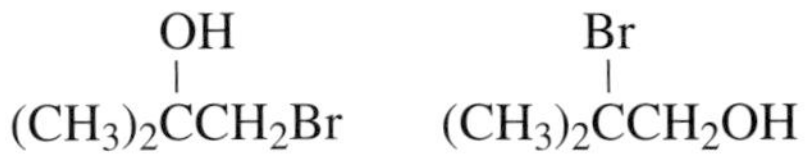

$$\underset{\textbf{1.9} \;\square}{\overset{\overset{\displaystyle OH}{|}}{(CH_3)_2CCH_2Br}} \qquad \underset{\textbf{1.10} \;\square}{\overset{\overset{\displaystyle Br}{|}}{(CH_3)_2CCH_2OH}}$$

In practice, these reactions obey the mechanistic Markovnikov rule, even though the intermediate is a bromonium ion rather than a carbocation. During attack of the nucleophile on the bromonium ion, the carbon–bromine bond breaks in advance of the carbon nucleophile bond being made, leading to a build up of positive charge on the electrophilic carbon in the transition state. Thus, the more alkyl groups that are attached to this carbon, the greater the spread of charge, the lower the energy of the transition state, and the faster the reaction, such that this isomer is the predominant product. Put simply, to predict the predominant product, we must decide which of the two *potential* carbocation intermediates is the more stable.

● Which product, **1.9** or **1.10**, will predominate in Reaction 1.35?

○ Compound **1.9** will predominate, because the potential intermediate carbocation leading to this product is tertiary:

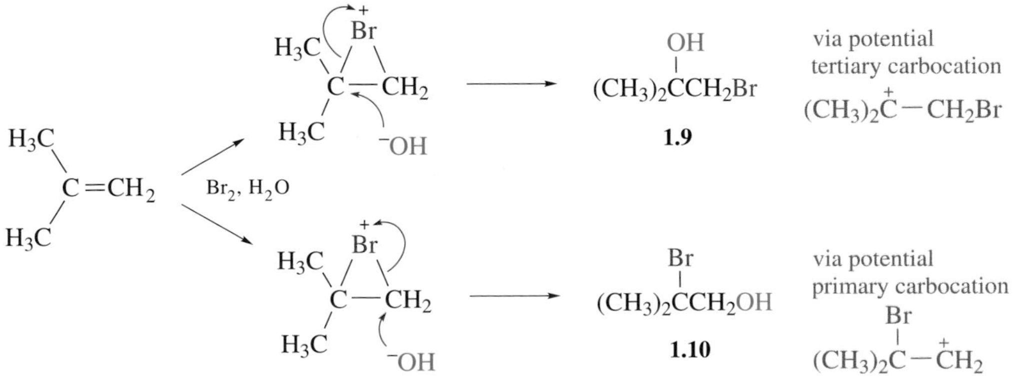

SCHEME 1.7

So, in addition reactions involving the bromonium ion, the electrophile (bromine) ends up on the carbon atom with the most hydrogen atoms attached to it. Of course, this is only apparent when bromide ion is not the nucleophile.

An alternative way of making bromohydrins is to use hypobromous acid, HOBr, which reacts in the same way as bromine. Oxygen is more electronegative than bromine, so the oxygen–bromine bond is polarized $\overset{\delta-}{HO}-\overset{\delta+}{Br}$. The bromine therefore acts as the electrophilic centre, leading again to the formation of a bromonium ion intermediate:

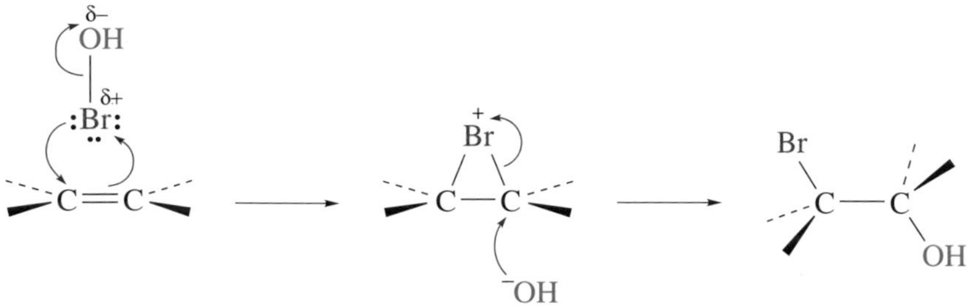

$$\text{(1.36)}$$

The hydroxide ion then attacks the bromonium ion from the opposite side to the bromine 'leaving group'.

Hypobromous acid can conveniently be made by treating N-bromoethanamide (N-bromoacetamide) with water in acid conditions:

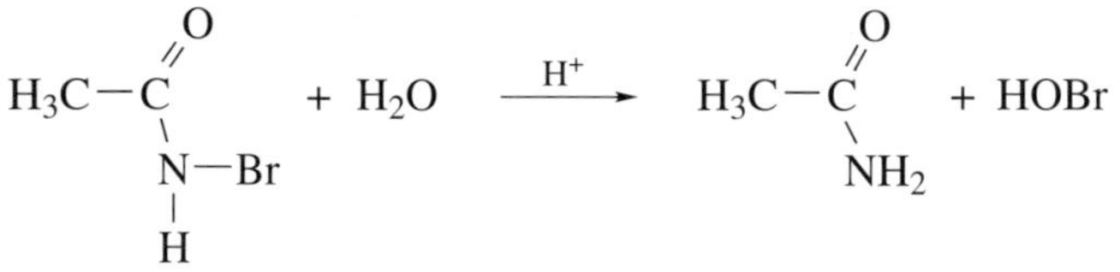

$$\text{(1.37)}$$

N-bromoethanamide

Which bromohydrin will be formed in the following reaction?

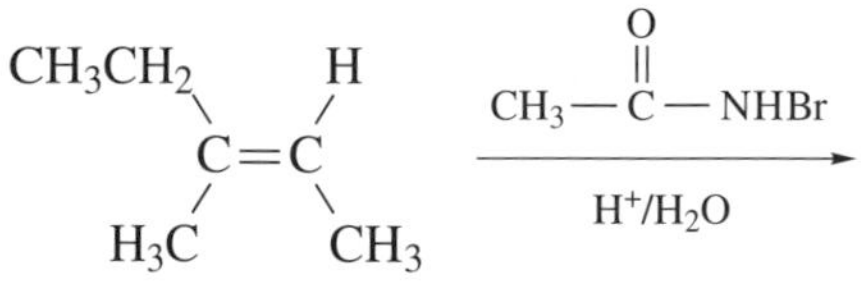

$$\text{(1.38)}$$

N-Bromoethanamide decomposes in water to give hypobromous acid, which reacts with the alkene to give a bromonium ion:

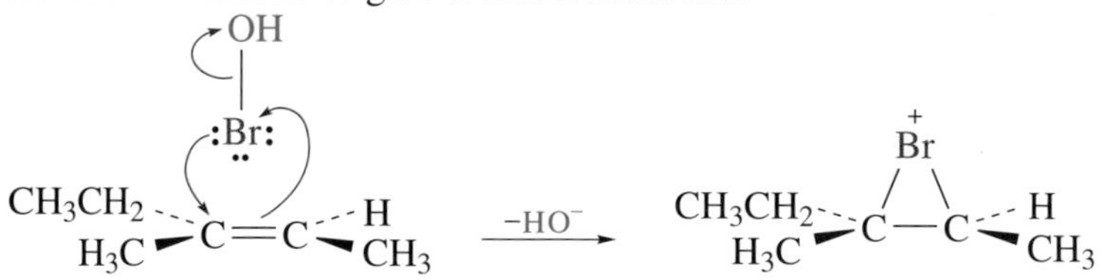

$$\text{(1.39)}$$

Since tertiary carbocations are more stable than secondary ones, the mechanistic Markovnikov rule predicts that attack of HO^- will be more likely to occur at the more-substituted carbon atom of the ring:

$$\text{(1.40)}$$

In fact, a racemic mixture is formed, because attack of HOBr from above or below the plane of the alkene is equally likely:

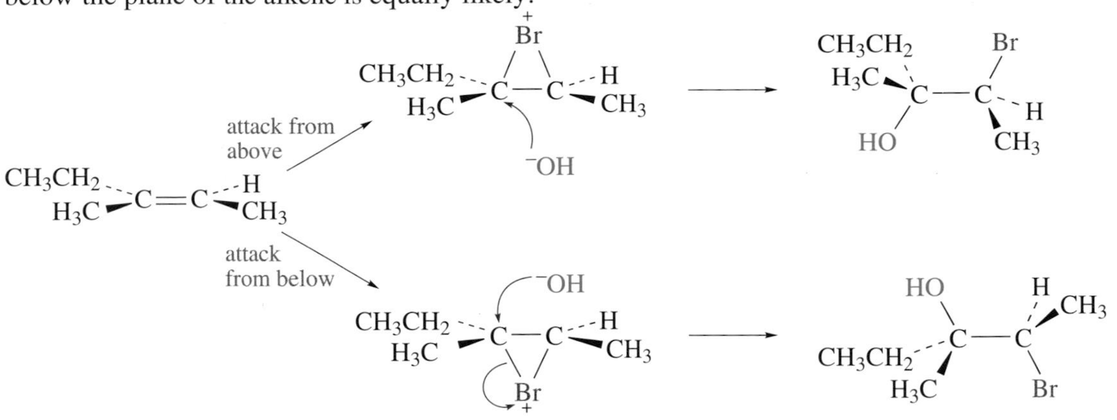

SCHEME 1.8

One important use of bromohydrins is in the formation of oxiranes (epoxides). These are compounds that contain two carbons and an oxygen in a three-membered ring. Essentially they are cyclic ethers, which because three atoms are constrained in a three-membered ring are more reactive than conventional, non-cyclic ethers. Treatment of the bromohydrin with base gives a small amount of alkoxide, which can undergo an internal S_N2 reaction.

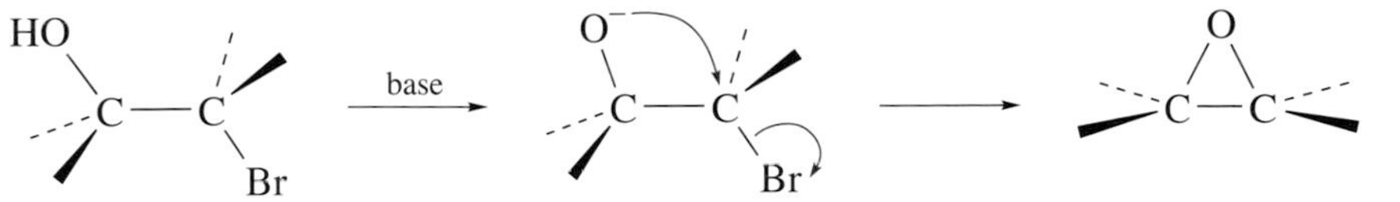

(1.41)

⬤ What will be the product if **1.11** is treated with base?

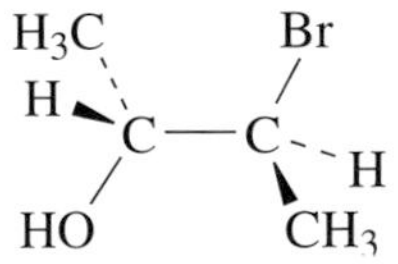

⬤ In S_N2 reactions the nucleophile approaches from the opposite side of the molecule to the leaving group. Hence the product will be as shown in Reaction 1.42:

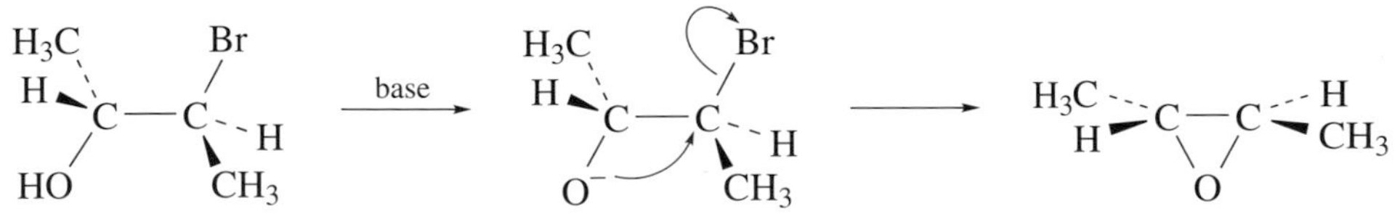

1.11

(1.42)

As we shall see in Part 3, oxiranes are very useful compounds, because by reacting with a nucleophile they create two adjacent functional groups. They also feature as a key structural element of epoxy resins (Box 1.5).

1.4 Summary of Section 1

1 In electrophilic addition reactions, the electrons of the double bond act as the nucleophile; an electrophilic reagent provides the other component of the reaction.

2 HCl, HBr, HI, and other strong acids such as H_2SO_4, provide protons that are sufficiently electrophilic to add to a double bond in the first step of an electrophilic addition reaction. The Markovnikov rule states that, with an unsymmetrical alkene, the proton adds to the carbon atom with the greater number of hydrogens attached to it. This can be rationalized because the major product arises from the more-stable carbocation intermediate. The carbocation then reacts with an anion (the nucleophile), to give the reaction product.

BOX 1.5 Epoxy resins

The oxirane (epoxide) functional group is important in epoxy resins such as Araldite™ (Figure 1.5). This adhesive is a polymer (**1.14**), which is made by reacting the disodium salt of bisphenol A, **1.12**, with an oxirane called epichlorohydrin, **1.13**. The two O⁻ groups on bisphenol A cause a nucleophilic substitution at the epoxy and chloro centres of the epichlorohydrin.

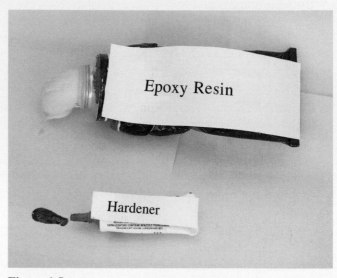

Figure 1.5
The two components of a typical epoxy resin glue.

disodium salt of bisphenol A, **1.12**

epichlorohydrin, **1.13**

1.14

hardener, **1.15**

To set the resin, the adhesive is mixed with a hardener, which is a polymer containing many amine groups, **1.15**. When mixed, the amino groups of the hardener undergo nucleophilic substitution at the epoxide groups of the adhesive to give a rigid network of cross-linked polymers. The final step of the reaction involves proton transfer from nitrogen to oxygen.

$$\text{(1.43)}$$

3   Bromine and chlorine molecules are polarized in the presence of an alkene, and so provide an electrophilic halogen atom to initiate the reaction. The intermediate from this reaction is a cyclic halonium (bromonium or chloronium) ion, and so the alkene substituents retain their original stereochemical relationship. Nucleophilic attack on the halonium ion occurs from the side remote from the halogen atom, and so the process is an *anti*-addition overall.

4   The intermediate halonium ion can be intercepted by nucleophiles other than bromide ion. If the reaction is carried out in methanol or water, 1,2-bromomethoxy compounds or 1,2-bromoalcohols, respectively, are produced; where appropriate, the major product is formed from the more-stable potential carbocation intermediate (that is, the mechanistic Markovnikov rule is obeyed).

5   Treatment of bromohydrins with base gives oxiranes.

QUESTION 1.1 ⌨

Predict the products of the following addition reactions. If two products are possible, say which one you think will predominate.

(a) $C_6H_5CH{=}CHC_6H_5 + HBr$

(b) $C_6H_5CH_2CH{=}CHCH_3 + HBr$

(c) 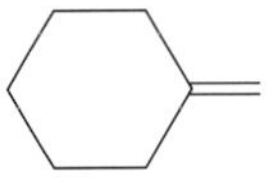 $={=}CH_2 + H_2O$ (plus acid catalyst)

QUESTION 1.2 ⌨

Give the products of the following reactions, indicating the stereochemistry where it is relevant:

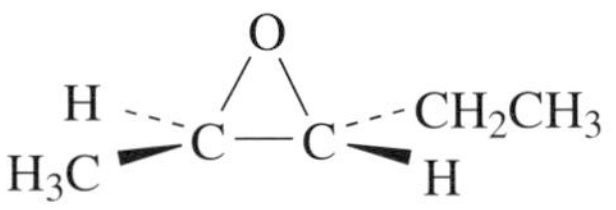

(a) $(CH_3)_2C{=}C(CH_3)_2 + Cl_2$ in tetrachloromethane

(b) $(C_6H_5)_2C{=}CH_2 + Br_2$ in methanol

(c) 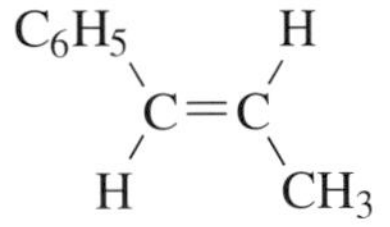 $+ Cl_2$ in tetrachloromethane

QUESTION 1.3 ⌨

What are the reactants needed to prepare the compounds **1.16–1.18**? (Ensure that the required stereochemistry will be obtained.)

meso

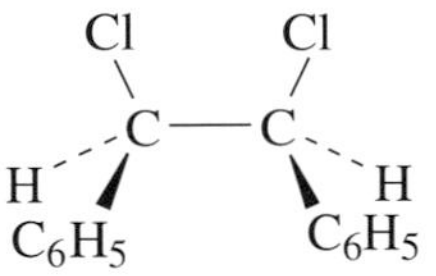

(i) **1.16** ⌨   (ii) **1.17** ⌨   (iii) **1.18** ⌨

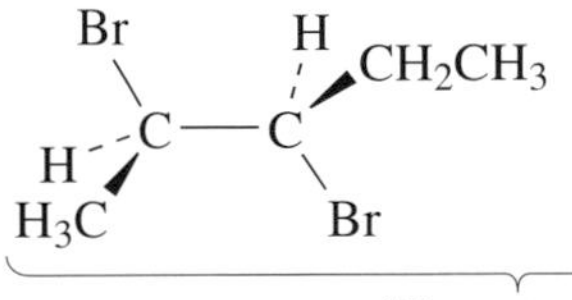

OTHER USEFUL ADDITION REACTIONS

2

The addition reactions we have studied so far all proceed by the electrophilic addition mechanism. There are several other useful addition reactions that proceed by a variety of mechanisms. We don't have space to discuss them all, but we shall mention one important example.

2.1 *syn*-Additions; hydrogenation

In certain addition reactions, both the new atoms or groups become attached to the same side of the alkene; that is, **syn-addition** is said to take place. We shall deal with only one of these reaction types, namely the hydrogenation of alkenes to alkanes, which is a reduction reaction.

Three examples are shown, in Reactions 2.1, 2.2 and 2.3:

$$CH_3CH{=}CH_2 \xrightarrow{\text{H}_2/\text{Ni}} CH_3CH_2CH_3 \qquad (2.1)$$

$$(2.2)$$

$$(2.3)$$

Alkenes react readily with hydrogen only in the presence of a metal catalyst: platinum, palladium and nickel are especially effective. The actual mechanism of such hydrogenation reactions is complex, but it appears to involve adsorption of the hydrogen on to the metal surface (the first step in Figure 2.1). Subsequent reaction with the alkene (which may also be adsorbed on to the surface of the catalyst) occurs, followed by desorption of the hydrogenated product. As both hydrogens are delivered from the metal surface, the predominance of the *syn*-addition product is not surprising. A practical application of *syn*-addition is discussed in Box 2.1 (p. 30).

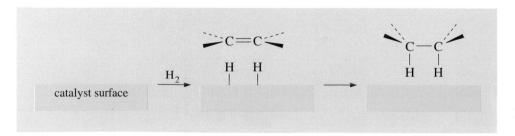

Figure 2.1
Schematic view of the catalytic hydrogenation of an alkene to an alkane.

Alkynes can also be hydrogenated, and, when the normal hydrogenation catalysts (platinum, palladium, nickel) are employed, the product is usually an alkane rather than an alkene:

$$R^1C{\equiv}CR^2 \xrightarrow{H_2} R^1CH{=}CHR^2 \xrightarrow{H_2} R^1CH_2CH_2R^2 \tag{2.4}$$

However, by using special deactivated catalysts, the hydrogenation reaction can be slowed down, and the alkene can be isolated. These catalysts are often referred to as 'poisoned' catalysts, and the one most frequently used is called Lindlar's catalyst. This consists of palladium on barium sulfate; it is 'poisoned' by the addition of a small amount of quinoline (**2.1**), which slows down the hydrogenation.

2.1

⬤ Do you think that the *cis* or the *trans* alkene will predominate from the hydrogenation of an alkyne?

⬤ Because the mechanism of hydrogenation involves the delivery of both hydrogen atoms from the metal surface to the same side of the alkyne, the *cis* alkene is usually formed to the exclusion of the *trans* isomer; for example

$$C_2H_5C{\equiv}CC_2H_5 \xrightarrow[\text{catalyst}]{\text{Lindlar's}} \underset{H \qquad H}{\overset{C_2H_5 \qquad C_2H_5}{\diagdown C{=}C \diagup}} \tag{2.5}$$

2.2 Summary of Section 2

1 Alkenes react readily with hydrogen in the presence of a transition-metal catalyst. This is a *syn*-addition.

2 Alkynes are hydrogenated in the presence of a metal catalyst to give alkanes. With deactivated (poisoned) catalysts such as the Lindlar type, alkynes give *cis* alkenes.

QUESTION 2.1 ⌨

What are the reagents needed to achieve the following conversions?

(a) $C_6H_5C{\equiv}CC_6H_5 \longrightarrow \underset{H \qquad H}{\overset{C_6H_5 \qquad C_6H_5}{\diagdown C{=}C \diagup}}$

(b) $C_6H_5C{\equiv}CC_6H_5 \longrightarrow C_6H_5CH_2CH_2C_6H_5$

BOX 2.1 Hydrogenation of fats — the manufacture of margarine

Margarine and butter are emulsions formed between fats and water (Figure 2.2). Fats are known as triglycerides, **2.2**. They are formed by the reaction of three fatty acid molecules, **2.3**, and glycerol, **2.4**, which contains three —OH groups:

$$R^1 \overset{\overset{\displaystyle O}{\|}}{C} OH \quad + \quad \begin{array}{c} CH_2-OH \\ | \\ CH-OH \\ | \\ CH_2-OH \end{array} \quad \longrightarrow \quad \begin{array}{c} CH_2-O-\overset{\overset{\displaystyle O}{\|}}{C}-R^1 \\ | \\ CH-OH \\ | \\ CH_2-OH \end{array} \quad + \quad H_2O \quad (2.6)$$

2.3 　　　　**2.4**　　　　　a monoglyceride

$$R^2 \overset{\overset{\displaystyle O}{\|}}{C} OH \quad + \quad \begin{array}{c} CH_2-O-\overset{\overset{\displaystyle O}{\|}}{C}-R^1 \\ | \\ CH-OH \\ | \\ CH_2-OH \end{array} \quad \longrightarrow \quad \begin{array}{c} CH_2-O-\overset{\overset{\displaystyle O}{\|}}{C}-R^1 \\ | \\ CH-O-\overset{\overset{\displaystyle O}{\|}}{C}-R^2 \\ | \\ CH_2-OH \end{array} \quad + \quad H_2O \quad (2.7)$$

a diglyceride

$$R^3 \overset{\overset{\displaystyle O}{\|}}{C} OH \quad + \quad \begin{array}{c} CH_2-O-\overset{\overset{\displaystyle O}{\|}}{C}-R^1 \\ | \\ CH-O-\overset{\overset{\displaystyle O}{\|}}{C}-R^2 \\ | \\ CH_2-OH \end{array} \quad \longrightarrow \quad \begin{array}{c} CH_2-O-\overset{\overset{\displaystyle O}{\|}}{C}-R^1 \\ | \\ CH-O-\overset{\overset{\displaystyle O}{\|}}{C}-R^2 \\ | \\ CH_2-O-\overset{\overset{\displaystyle O}{\|}}{C}-R^3 \end{array} \quad + \quad H_2O \quad (2.8)$$

a triglyceride, **2.2**

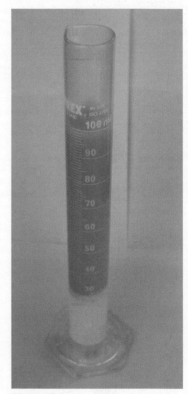

Figure 2.2
Margarine is an emulsion of water in a fat; when heated, the fat melts and the two phases separate.

The fats from animals (for example, **2.5**) are usually saturated; that is, the fatty acids contain no carbon–carbon double bonds. The oils obtained from plants are also triglycerides, but, the fatty acids they are derived from are usually unsaturated; see, for example, **2.6**.

$$\begin{array}{c} CH_2-O-\overset{\overset{\displaystyle O}{\|}}{C}-(CH_2)_{16}-CH_3 \\ | \\ CH-O-\overset{\overset{\displaystyle O}{\|}}{C}-(CH_2)_{16}-CH_3 \\ | \\ CH_2-O-\overset{\overset{\displaystyle O}{\|}}{C}-(CH_2)_{16}-CH_3 \end{array}$$

a typical animal triglyceride, **2.5**

$$\begin{array}{c} CH_2-O-\overset{\overset{\displaystyle O}{\|}}{C}-(CH_2)_7-CH=CH-CH_2-CH=CH-(CH_2)_4-CH_3 \\ | \\ CH-O-\overset{\overset{\displaystyle O}{\|}}{C}-(CH_2)_7-CH=CH-CH_2-CH=CH-(CH_2)_4-CH_3 \\ | \\ CH_2-O-\overset{\overset{\displaystyle O}{\|}}{C}-(CH_2)_7-CH=CH-CH_2-CH=CH-(CH_2)_4-CH_3 \end{array}$$

a typical vegetable triglyceride, **2.6** 💻

The double bonds usually have the *cis* configuration, such that the chains of the triglycerides do not pack well together. This results in triglycerides from plants having low melting temperatures — they are oils at room temperature — so they are not very useful for making a solid spread. However, hydrogenation of some of the double bonds in vegetable oils leads to less unsaturation, converting the oils into solid fats that can be used to make a solid spread. In a typical process the vegetable oil is mixed with finely divided nickel, and hydrogen is pumped through the mixture. The reaction takes place above 170 °C and at high pressures (200–700 kPa). Although the reaction needs heating at first,

the reaction is exothermic so further heating is not required. The hot oil is filtered to remove the nickel, and left to cool down and solidify. The 'chemical' nature of margarine production is highlighted in Figure 2.3, which is from an advert to promote the 'more natural' butter.

Saturated fats are thought to be unhealthy and lead to heart disease, whereas margarine, which still contains reasonable amounts of unsaturation, is thought to be more healthy. However, recently there has been concern over the formation of *trans* fatty acid esters in the hydro-genation process. This arises from isomerization of the *cis* fatty acid ester to the *trans* form on the surface of the metal.

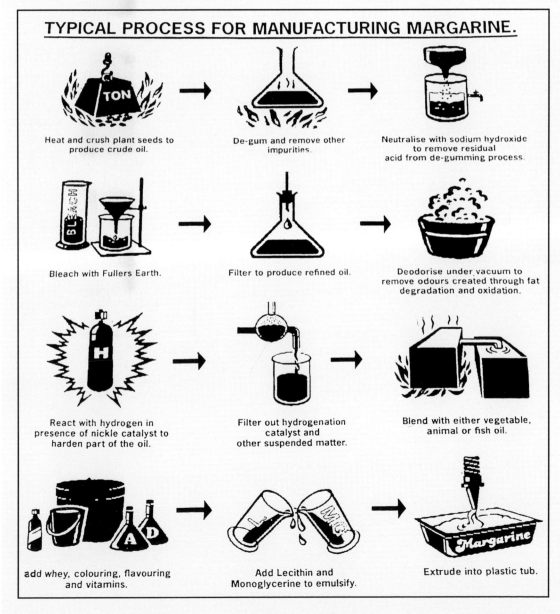

TYPICAL PROCESS FOR MANUFACTURING MARGARINE.

Heat and crush plant seeds to produce crude oil.

De-gum and remove other impurities.

Neutralise with sodium hydroxide to remove residual acid from de-gumming process.

Bleach with Fullers Earth.

Filter to produce refined oil.

Deodorise under vacuum to remove odours created through fat degradation and oxidation.

React with hydrogen in presence of nickle catalyst to harden part of the oil.

Filter out hydrogenation catalyst and other suspended matter.

Blend with either vegetable, animal or fish oil.

add whey, colouring, flavouring and vitamins.

Add Lecithin and Monoglycerine to emulsify.

Extrude into plastic tub.

Figure 2.3
Illustration from an advert to promote butter!

LEARNING OUTCOMES

Now that you have completed *Alkenes and Aromatics*: *Addition—pathways and products*, you should be able to do the following things:

1   Recognize valid definitions of, and use in a correct context, the terms, concepts and principles in the following Table. (All Questions)

List of scientific terms, concepts and principles introduced in *Addition—pathways and products*

Term	Page number
addition reactions	11
bromohydrin	23
cyclic bromonium ion	20
Markovnikov's rule	15
mechanistic Markovnikov rule	16
syn-addition	28

2   Given the reactants in an addition reaction: predict the structure of the product, indicating its stereochemistry, and, where more than one addition product can be formed, predict which one will predominate. (Questions 1.1 and 1.2)

3   Specify the reagents and conditions needed to form a given product by an addition reaction. (Questions 1.2, 1.3 and 2.1)

QUESTIONS: ANSWERS AND COMMENTS

QUESTION 1.1 (Learning Outcome 2)

(a) The product would be compound **Q.1**. Since one of the central carbon atoms is chiral in **Q.1**, both enantiomers will be formed in equal amounts; that is, a racemic mixture would be formed.

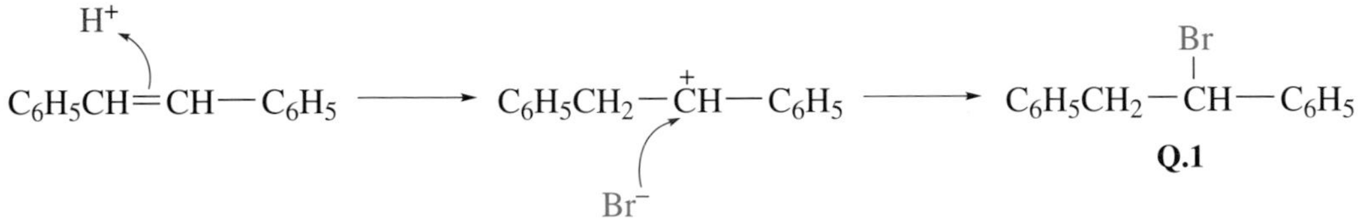

$$C_6H_5CH=CH-C_6H_5 \xrightarrow{\;H^+\;} C_6H_5CH_2-\overset{+}{C}H-C_6H_5 \xrightarrow{\;Br^-\;} C_6H_5CH_2-\overset{\displaystyle Br}{\underset{}{C}}H-C_6H_5 \qquad (Q.1)$$

Q.1

(b) Both carbocation intermediates are secondary, and a mixture of both products, **Q.2** and **Q.3**, will probably be formed as racemic mixtures.

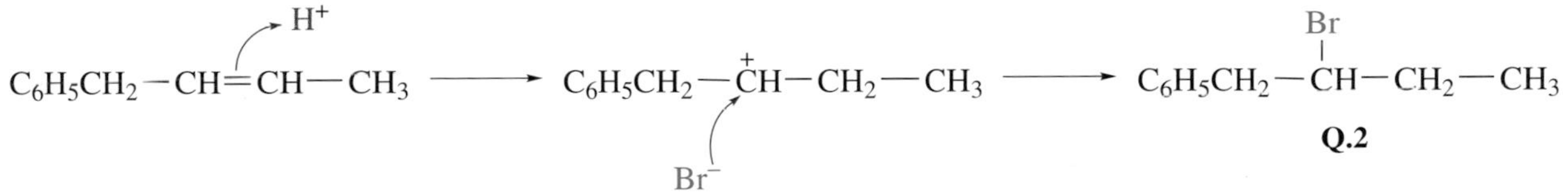

$$C_6H_5CH_2-CH=CH-CH_3 \xrightarrow{\;H^+\;} C_6H_5CH_2-\overset{+}{C}H-CH_2-CH_3 \xrightarrow{\;Br^-\;} C_6H_5CH_2-\overset{Br}{C}H-CH_2-CH_3 \quad (Q.2)$$

Q.2

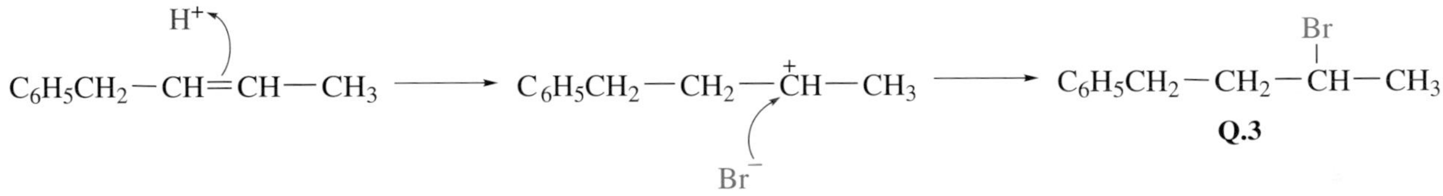

$$C_6H_5CH_2-CH=CH-CH_3 \xrightarrow{\;H^+\;} C_6H_5CH_2-CH_2-\overset{+}{C}H-CH_3 \xrightarrow{\;Br^-\;} C_6H_5CH_2-CH_2-\overset{Br}{C}H-CH_3 \quad (Q.3)$$

Q.3

(c) Protonation will occur to give the tertiary carbocation, which will give rise to the tertiary alcohol **Q.4**.

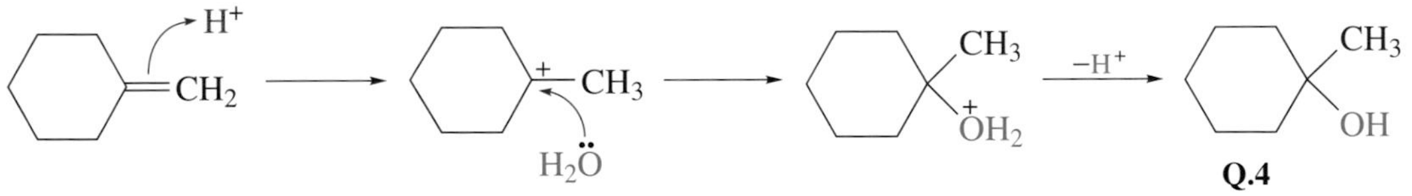

(Q.4)

Q.4

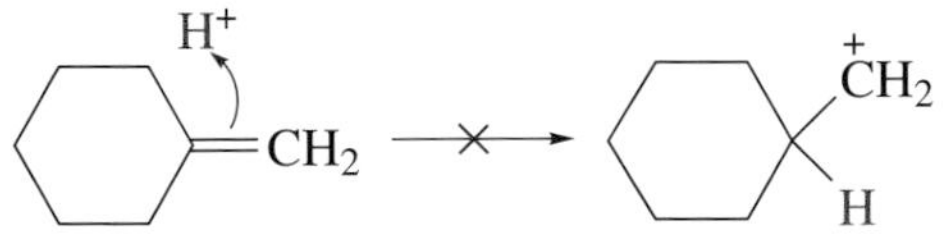

the primary carbocation is less favoured

QUESTION 1.2 (Learning Outcomes 2 and 3)

(a) The product will be $(CH_3)_2CClCCl(CH_3)_2$.

(b) The intermediate bromonium ion will be intercepted by methanol and Br^-. Nucleophilic attack will occur preferentially at the site of the more-stable potential carbocation, and two major products will be formed: $(C_6H_5)_2C(OCH_3)CH_2Br$ and $(C_6H_5)_2CBrCH_2Br$, whose relative proportions will depend on the concentration of bromine.

(c) Initial attack of the chlorine can take place from above or below the plane of the double bond. When the chloronium ion is formed, the Cl^- will attack the carbon attached to the phenyl group, since this would be the site of the more-stable carbocation (stabilization of a carbocation by a phenyl group is equivalent to that resulting from two alkyl groups). The two products formed from this reaction, **Q.5** and **Q.6**, are enantiomers.

attack of chlorine from above

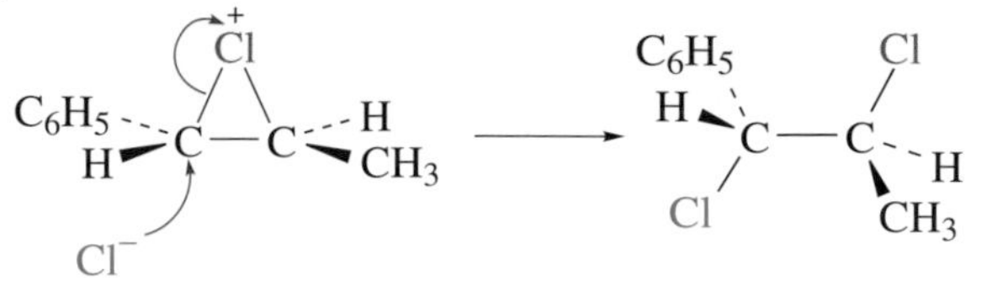

(Q.5)

Q.5

attack of chlorine from below

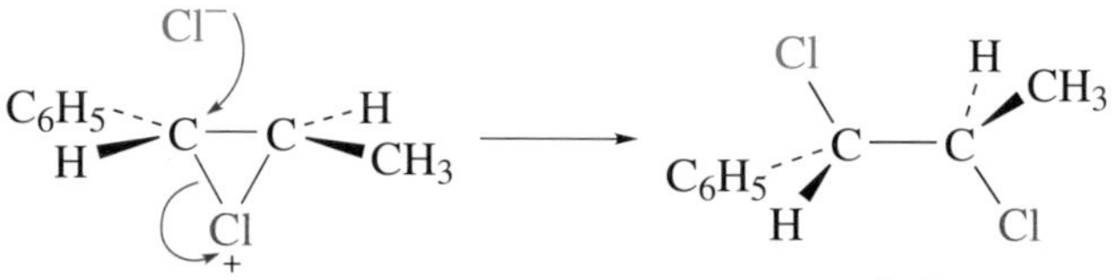

(Q.6)

Q.6

QUESTION 1.3 *(Learning Outcome 3)*

In all three cases, it is assumed that the product is formed by an *anti*-addition process. The simplest way of tackling this problem is to make a model or look at the WebLab ViewerLite image, and then rotate about the central carbon–carbon bond until the two halogen atoms are in an antiperiplanar relationship. This then highlights the relationship between the groups in the required alkene reactant.

(i) 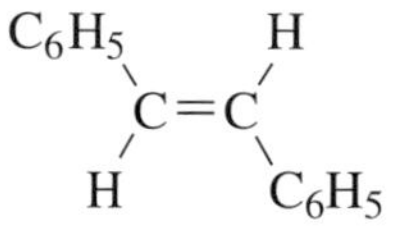 + Cl_2 gives **1.16**

(ii) 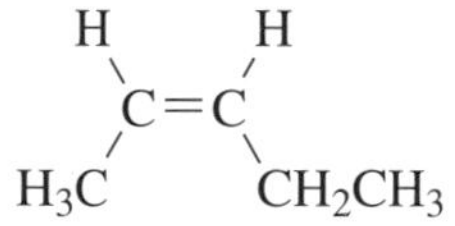 + Br_2 gives **1.17**

(iii) 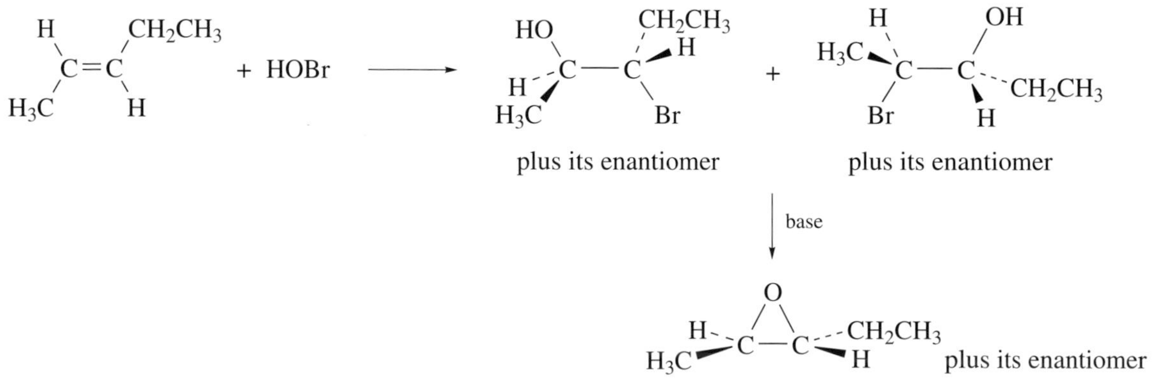
+ HOBr
plus its enantiomer
+
plus its enantiomer
base
plus its enantiomer

Q.7

In the third reaction, because both alkene carbon atoms give secondary carbocations, a mixture of bromohydrins will be formed. However, treatment with base gives just one oxirane (epoxide, **Q.7**).

QUESTION 2.1 (*Learning Outcome 3*)

(a) This is a catalytic hydrogenation reaction, stopping at the alkene stage. A poisoned catalyst such as the Lindlar catalyst is therefore needed.

(b) This is again a catalytic hydrogenation reaction, but in this case a 'normal' catalyst such as palladium, platinum or nickel can be used because the desired product is an alkane derivative.

FURTHER READING

1 M. Mortimer and P. G. Taylor (eds), *Chemical Kinetics and Mechanism*, The Open University and the Royal Society of Chemistry (2002).

ACKNOWLEDGEMENTS

Grateful acknowledgement is made to the following sources for permission to reproduce material in this book:

Figure 1.3: Courtesy of University of Pennsylvania Library; *Figure 2.3*: The Butter Council.

Every effort has been made to trace all the copyright owners, but if any has been inadvertently overlooked, the publishers will be pleased to make the necessary arrangements at the first opportunity.

Part 2

Aromatic Compounds

edited by Peter Taylor

based on Aromatic compounds
by Peter Morrod and Alan Bassindale (1991)

INTRODUCTION

One of the major preoccupations of eighteenth- and nineteenth-century technology in the United Kingdom was the utilization of one of the country's main natural resources, namely coal. At first, the emphasis was placed on the production of coke for iron smelting; all the other products, apart from a tar residue, were gaseous and were discarded into the atmosphere. It was not until the end of the eighteenth century that work was begun to explore possible uses for both the gaseous products and the tar. The observation that coal gas was flammable, 'burning with a light of great illuminating power', was to have enormous social consequences. By 1840 the development of gas lighting had progressed to the point where coal gas was being used routinely for the illumination of factories, schools and many private homes. Theatres and other public places quickly took advantage of gas lighting. The working day was extended, evening classes were established, and a new type of worker appeared — the lamplighter. There were problems though, as Figure 1.1 illustrates.

Figure 1.1
Cartoon by Richard Dighton, 1822. It is one of a series of 'London Nuisances', and satirizes the early use of coal gas, showing an explosion in the window of a chemist's shop.

Apart from the risk of explosions, another serious concern was the accumulation of large amounts of coal tar, which no one seemed to know what to do with. It was not unusual to find evidence of surplus tar having strangely found its way into rivers! In short, the country had a pollution problem on its hands. Attempts were made to extract more gas from coal tar, but these were only moderately successful.

The Royal Navy experimented with using the residue for tarring its ropes, but abandoned this within two years because of the consequent health hazards to sailors. We now know that coal tar contains many substances capable of giving rise to skin cancer after prolonged contact. By far the most successful application of coal tar at this time was its use for preserving wooden railway sleepers. Enormous amounts of wood were used to support the rails and, without the use of an effective preservative, the sleepers would have had to be replaced very frequently.

One of the Advantages of GAS over OIL.

Very few people in 1850 could have expected that coal tar would provide the raw material for a striking new industrial development, and no one anticipated that within this black, sticky, often evil-smelling substance some 250 different compounds would be identified, many of which would be found to have useful applications. The birth of the organic chemical industry was based on the careful distillation of coal tar (Figure 1.2). Coal tar is a mixture of many different compounds, and these distil off in an order determined by their boiling temperatures. The various fractions that were obtained contained a mixture of compounds boiling between two temperatures. Further fractional distillation of these mixtures led to the isolation of individual compounds. About one kilogram of benzene can be obtained from a thousand kilograms of coal. Although this is not a very high percentage yield, so much coal is now being coked globally that the annual production of benzene from the resulting coal tar is enormous. Even in the nineteenth century, there was plenty of coal tar just waiting to be used; as indicated in Box 1.1, some of it found application in surgical procedures.

It would be hard to exaggerate the importance to the chemical industry, and to our present economy, of the large-scale production of benzene and its derivatives. Just as the alkanes obtained from petroleum (see the Industrial Organic Chemistry Case Study at the end of this Book) are ultimately the source of nearly all our aliphatic compounds, so benzene, and compounds made from benzene, are ultimately the source of nearly all our aromatic compounds. When a chemist wishes to make a complicated aromatic compound, whether in the laboratory or on an industrial scale, a benzene ring is not built up from non-cyclic reactants: a simpler benzene derivative is modified to obtain the required structure.

Thus, benzene is the parent hydrocarbon of a very large class of organic compounds, which have come to be known as **aromatic compounds**. The origin of

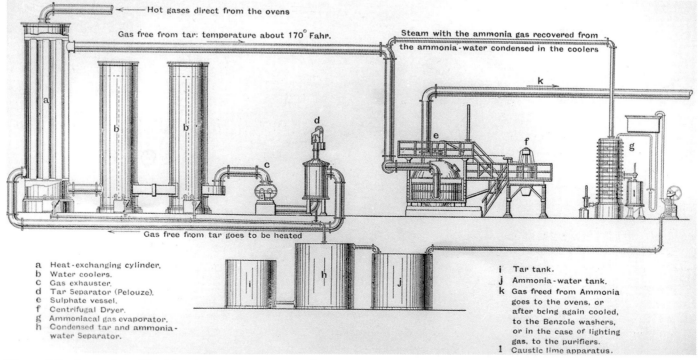

a   Heat-exchanging cylinder.
b   Water coolers.
c   Gas exhauster.
d   Tar Separator (Pelouze).
e   Sulphate vessel.
f   Centrifugal Dryer.
g   Ammoniacal gas evaporator.
h   Condensed tar and ammonia-
    water Separator.

i   Tar tank.
j   Ammonia-water tank.
k   Gas freed from Ammonia
    goes to the ovens, or
    after being again cooled,
    to the Benzole washers,
    or in the case of lighting
    gas, to the purifiers.
l   Caustic lime apparatus.

Figure 1.2   Separation of liquid and gaseous components from coal.

this classification is simple, but its use can be misleading. The word 'aromatic' (from the Greek word that means 'fragrant smell') stems from the fact that many of these compounds that were isolated from natural sources had one obvious thing in common — a pleasant odour; notable examples are vanilla, oil of wintergreen, cinnamon and aniseed. Later, when it was found that these compounds were also related to benzene and that this was their parent compound, the term 'aromatic' was adopted for benzene and all its derivatives. Given this historical explanation, it should be pointed out that many aromatic compounds do not smell pleasant, even to an organic chemist! Rather, the term should be thought of as indicating that a compound is related to benzene.

BOX 1.1 Surgical advances using coal tar

In the 1850s, two pharmacists in France, Ferdinand LeBeuf and Jules Lemaire demonstrated that coal tar possessed antiseptic properties. Meanwhile, in Manchester, the industrial chemist F. C. Calvert was interested in carbolic acid because it was a useful intermediate from coal tar for use in the dyestuffs industry. Calvert sold most of his carbolic acid to France, and suggested to the French Academy of Science that the disinfectant properties of coal tar arose from carbolic acid. Lemaire confirmed that this was true. About this time, Joseph Lister was looking for substances that would kill bacteria and so prevent infection during surgery. He investigated the use of carbolic acid for this purpose; in due course he found that it could be utilized as an antiseptic during surgery (Figure 1.3).

Carbolic acid, or phenol (**1.1**) as it is more commonly known, contains a benzene ring and an OH group. As well as being an antiseptic, phenol is corrosive and not the kind of thing you would want to put on your skin. However, compounds with other alkyl groups on the ring are less harmful. For example, thymol (**1.2**), which is a naturally occurring compound found in thyme, is used in mouthwashes and gargles.

Such phenolic compounds are present in coal tar soap (Figure 1.4), and contribute to its characteristic smell.

1.1   **1.2**

Figure 1.4
Coal tar soap.

Figure 1.3 In 1965, Britain celebrated the centenary of Lister's contribution to the development of antiseptic surgery by issuing two stamps; the one shilling stamp shows a rather odd representation of the molecular structure of phenol, the compound known to Lister as 'carbolic acid'.

THE STRUCTURE AND STABILITY OF BENZENE

2

The elucidation of the structure of benzene, and an understanding of its peculiar stability and chemical reactions, have provided some of the landmarks of organic chemistry. The correct structure for benzene was first proposed in 1865 by Friedrich August Kekulé (Box 2.1), and aromatic chemistry has remained an area of particularly active research ever since.

BOX 2.1 Friedrich August Kekulé von Stradonitz

Kekulé (Figure 2.1), who was born on 7 September 1829 in Darmstadt, Germany, was descended from a Bohemian noble family. In the winter of 1847, Kekulé enrolled at the University of Giessen to study architecture, but was so inspired by his chemistry teacher, Justus von Liebig, that he changed to chemistry, graduating in 1851. He then went to Paris to study for his Doctorate. In 1854–5 he worked as an assistant to J. Stenhouse at St Bartholomew's Hospital, London. He then moved to Heidelberg, subsequently becoming professor first at Ghent (1858) and then at Bonn (1867). He first married in 1862, but his wife died during childbirth. He married again in 1876, and had three more children. He also caught measles in 1876, and this affected his health for the remainder of his life.

Kekulé was apparently neither a particularly good practical chemist nor an inspiring teacher. However, he was a good theoretician and dreamer:

> During my stay in London I resided in Clapham Road….I frequently, however, spent my evenings with my friend Hugo Mueller….We talked of many things but most often of our beloved chemistry. One fine summer evening I was returning by the last bus, riding outside as usual, through the deserted streets of the city….I fell into a reverie, and lo, the atoms were gamboling before my eyes. Whenever, hitherto, these diminutive beings had appeared to me, they had always been in motion. Now, however, I saw how, frequently, two smaller atoms united to form a pair: how a larger one embraced the two smaller ones; how still larger ones kept hold of three or even four of the smaller: whilst the whole kept whirling in a giddy dance. I saw how the larger ones formed a chain, dragging the smaller ones after them but only at the ends of the chains….The cry of the conductor: 'Clapham Road' awakened me from my dreaming; but I spent a part of the night in putting on paper at least sketches of these dream forms.

This led to his proposals that carbon atoms with a valency of four can form chains with each other of any length or shape. He later dreamt the solution to the structure of benzene (1865):

> I turned my chair to the fire [after having worked on the problem for some time] and dozed. Again the atoms were gamboling before my eyes. This time the smaller groups kept modestly to the background. My mental eye, rendered more acute by repeated vision of this kind, could not distinguish larger structures, of manifold conformation; long rows, sometimes more closely fitted together; all twining and twisting in snakelike motion. But look! What was that? One of the snakes had seized hold of its own tail, and the form whirled mockingly before my eyes. As if by a flash of lightning I awoke…. Let us learn to dream, gentlemen.

Figure 2.1
Friedrich August Kekulé von Stradonitz 1829–1896.

The special problems associated with understanding the structure and properties of benzene stemmed from the knowledge that its molecular formula is C_6H_6. This formula requires that benzene contains four double bonds, or four rings, or two triple bonds, or, more likely, some combination of rings and multiple bonds. However, benzene does not easily undergo addition reactions, but it readily undergoes substitution reactions, in which an electrophile, X^+, leads to a C—X bond being formed, and a C—H bond being broken, with H^+ as a leaving group. Furthermore, only one substitution product, C_6H_5X, can be formed, requiring that all six hydrogen atoms are in chemically identical environments. To account for these, and other, observations, Kekulé proposed Structure **1.3** for benzene:

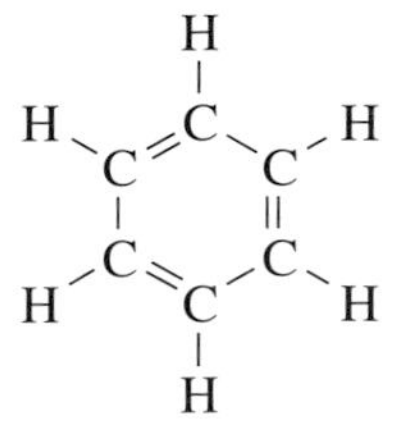

1.3

He also proposed (to account for some structural observations) that the double bonds were 'oscillating' between the two possible positions **1.3** and **1.4**:

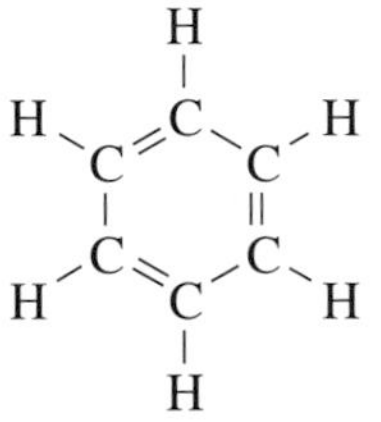

1.3   and   **1.4**

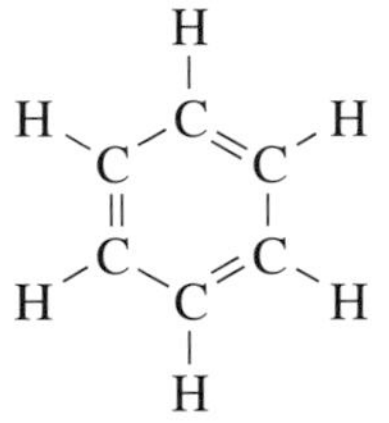

Modern techniques, such as X-ray crystallography have shown that all the carbon–carbon bonds in benzene are of exactly the same length, forming a regular hexagon with sides 140 pm long. The carbon–carbon bond length in benzene lies between that for a carbon–carbon single bond (154 pm) and a double bond (133 pm). The benzene molecule is perfectly planar, all bond angles are 120°, and the C—H bonds are all 108 pm long (**1.5**).

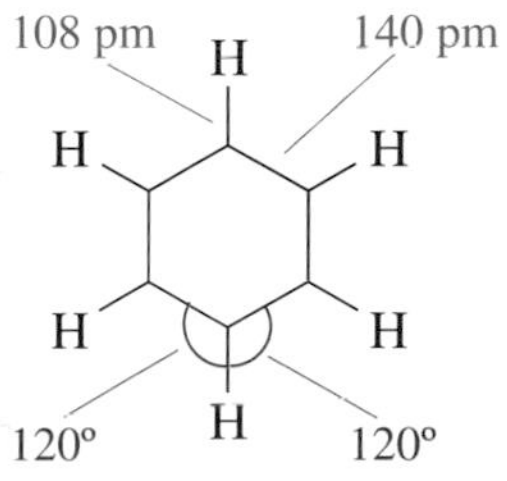

1.5

We no longer think that the bonds in benzene are oscillating, but the concept of resonance can be used to explain why all the carbon–carbon bonds in benzene have the same length. The two resonance forms of benzene contribute equally to the actual structure, so that each carbon–carbon bond is intermediate in nature between a single bond and a double bond; we say that the electrons are *delocalized* around the ring.

QUESTION 2.1

Write in your own words what is meant by the following representation:

1.6

From now on, for convenience, we shall draw benzene as a single resonance form — that is, as one of the two forms shown as **1.6**. Always remember that this is simply a convenient representation, and that in the real structure all the C—C bonds have equal bond lengths. Also, remember that a hydrogen atom is attached to each carbon atom.

Although the idea of resonance stabilization is useful in helping to understand the structure and properties of benzene, it is not quite sufficient. As you will see in detail in the next Section, benzene derivatives have a very strong tendency to maintain intact the six-electron π system (from the three double bonds). We cannot explore the concept of **aromaticity** in detail here, but it has been shown that planar, cyclic, π systems with $(4n + 2)$ π electrons ($n = 0, 1, 2, 3$, etc.) benefit from a considerable extra stabilization over molecules not containing these numbers of π electrons (that is, compared with non-aromatic systems). In benzene, this extra aromatic stabilization is about $260\,kJ\,mol^{-1}$.

● Before moving on to look at the chemistry of benzene, try to draw a simple picture of the way in which the p orbitals overlap to form the π system in benzene.

● In benzene, each carbon atom is bonded to three other atoms, and can be described as sp^2 hybridized (bond angles 120°). The molecule is therefore perfectly set up for a planar σ bond skeleton (**1.7**).

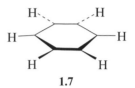

1.7

As shown in Figure 2.2a, the remaining p orbitals are perpendicular to the plane of the ring, and initially each has one electron associated with it.

The p orbitals can interact equally strongly with either neighbour, and the end result is a **delocalized p-electron system** above and below the plane of the benzene ring, as shown in Figure 2.2b.

2.1 Summary of Sections 1 and 2

1 At the beginning of the nineteenth century, coal tar was considered an unusable byproduct from the coking of coal, but by the end of the century it had proved to be important for the birth of a new industry in Britain: the organic chemical industry.

2 Distillation of coal tar provides numerous aromatic compounds, many of which are of industrial significance.

3 This Part of the Book is mainly concerned with the simplest aromatic compound — benzene.

4 Benzene can be represented by the following resonance forms:

5 All the carbon–carbon bonds in benzene are of equal length, as are all the carbon–hydrogen bonds. The bonds between the carbon atoms are intermediate in length between single and double carbon–carbon bonds. All bond angles are 120°.

QUESTION 2.2

One of the reasons why Kekulé suggested 'oscillating' single and double bonds for benzene, rather than a localized structure, was that there were only three disubstituted chlorobenzenes, $C_6H_4Cl_2$. Draw structures for each of the isomers of $C_6H_4Cl_2$, (a) assuming that all the carbon–carbon bonds are identical, and (b) assuming a localized bond structure. Use these structures to show why it was necessary to propose the delocalized structure.

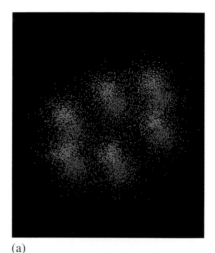

(a)

(b)

Figure 2.2
(a) Disposition of the p orbitals on the carbon atoms of benzene prior to π-bond formation; (b) π-electron distribution in benzene after overlap of the unhybridized p orbitals.

ELECTROPHILIC AROMATIC SUBSTITUTION REACTIONS

3

3.1 General principles

The most characteristic reaction of benzene and other aromatic systems is classed as a substitution reaction:

$$
\text{(benzene)} + XY \longrightarrow \text{(X-substituted benzene)} + HY \qquad (3.1)
$$

● Which bonds are being made and which are broken in this general reaction?

● Remember that each carbon atom in the benzene ring has one hydrogen atom attached to it (even though this is not shown in the structural formula), so a C—H bond is being broken and a new C—X bond is being made. Hence, the X group is *substituting* for a hydrogen atom in the molecule.

When we consider the mechanism, we shall discover that a cation, X^+, is usually involved in reacting with the benzene ring.

● Should the X^+ cation be classed as an electrophile or a nucleophile?

● It is an electrophile. The positive charge indicates that X^+ has a low electron density. Its chemical behaviour will be to seek electron-rich molecules with which to react, and this is exactly what is meant by an electrophilic reagent.

In summary, then, the general reaction that is characteristic of aromatic compounds is not just a substitution reaction, but, because the attacking reagent is an electrophile, it is an **electrophilic aromatic substitution reaction**. We need to understand why an electrophilic reagent does not *undergo addition* to any or all of the three equivalent double bonds that we can postulate to be present in the resonance forms of benzene. Consider the following (hypothetical) reaction scheme:

$$
\qquad (3.2)
$$

This is just an electrophilic addition of XY to one of the double bonds in the benzene ring. After studying the addition reactions in Part 1, you may well assume that what we have shown as the reaction product is entirely feasible. In practice, though, benzene (almost) never undergoes this kind of reaction. The reason is concerned with the particular stability of the cyclic arrangement of three double bonds alternating with three single bonds[*]. Let us look more closely at the mechanism of electrophilic aromatic substitution reactions. It is important to

[*] The molecular orbitals arising from such delocalization are discussed in *Molecular Modelling and Bonding*[1].

recognize that although benzene can undergo many substitution reactions, there really is only one general mechanism for all electrophilic aromatic substitution reactions.

There are usually three identifiable steps in any electrophilic aromatic substitution reaction:

1 Generation of the electrophile

A reagent XY either spontaneously generates X^+, or is induced to do so by another reagent or by a catalyst:

$$XY \longrightarrow X^+ + Y^- \tag{3.3}$$

2 Attack at the benzene ring by the electrophile

In this step, the benzene ring acts as a nucleophile in reacting with X^+. This is exactly what we would expect for the first step in an addition to a simple double bond:

$$\tag{3.4}$$

The resulting carbocation, which is strongly stabilized, is an intermediate in the overall reaction.

You should now attempt the following question.

QUESTION 3.1

How can the carbocation formed in electrophilic aromatic substitution be stabilized? Use drawings to illustrate your answer.

The carbocation intermediate can be drawn as either one of the three resonance forms shown in the answer to Question 3.1. We can represent the delocalized nature of this intermediate more succinctly as in Structure **3.1**. Note that the broken line starts on C-2 and ends on C-6. There is no double-bond character at C-1, since it is already connected to four atoms by single bonds.

3.1

3 Regeneration of the benzene aromatic ring by loss of H^+

It is the step shown in Equation 3.5 that is particularly characteristic of electrophilic aromatic substitution reactions:

$$+ \ YH \tag{3.5}$$

The driving force for this step is the re-formation of the very stable cyclic arrangement of three double bonds. This force is so strong that a reagent Y^- will invariably remove the proton as shown above, rather than add to one of the carbocation centres. In general, electrophilic aromatic substitution is favoured over addition, on both thermodynamic and kinetic grounds.

We can represent the course of a typical electrophilic aromatic substitution reaction on a reaction-coordinate diagram, as shown in Figure 3.1. The activation energy, E_a, for the reaction is usually that for the first step. So attack of X^+ on the benzene ring is generally the slow, rate-limiting step. In this first, slow, step the intermediate carbocation is formed. The decomposition of the intermediate is usually fast, since the proton is quite easily removed to re-form the aromatic ring. Finally, electrophilic aromatic substitution reactions usually, but not exclusively, take place with a reduction of the enthalpy of the system; that is, they are *exothermic*.

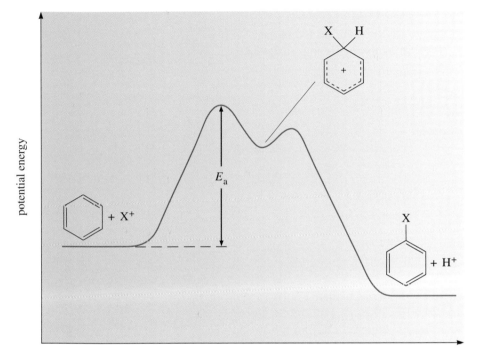

reaction coordinate

Figure 3.1
The reaction-coordinate diagram for a typical electrophilic aromatic substitution reaction.

● What do you expect the rate equation to be for an electrophilic aromatic substitution reaction between X^+ and benzene?

● As the first step is slow, the electrophile is involved in the rate-limiting step. Therefore, the rate equation is expected to be

$$J_{obs} = k[\text{benzene}][X^+] \tag{3.6}$$

An experimental rate equation of this form is found for most electrophilic aromatic substitution reactions.

3.1.1 Summary of Section 3.1

1   The substitution reactions discussed in this Section are electrophilic aromatic substitutions, as distinct from nucleophilic substitutions. In contrast, the electrophilic reactions we discussed in Part 1 were *addition* reactions.

2   Benzene does not undergo electrophilic addition reactions because the cyclic system of three double bonds alternating with three single bonds confers particular stability on products as well as on reactants.

3   Electrophilic substitution proceeds by way of a carbocation intermediate.

4   The general mechanism of electrophilic aromatic substitution reactions consists of two steps — a slow, rate-limiting step, followed by a fast step:

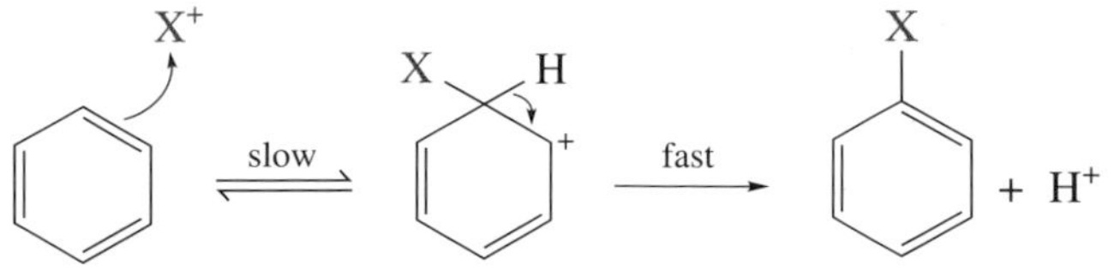

(3.7)

QUESTION 3.2

Draw the resonance forms of the intermediate carbocation formed in the following reactions of an electrophile (X^+) with methylbenzene (toluene, **3.2**):

(a) reaction at a position *ortho* to the CH_3 group of methylbenzene (that is, positions 2 or 6 on the ring);

(b) reaction at a *meta* position (positions 3 or 5);

(c) reaction at the *para* position (position 4).

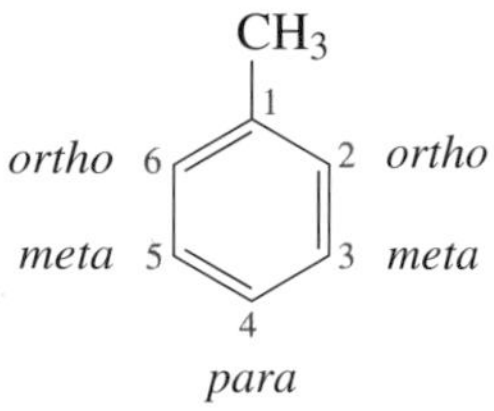

3.2

QUESTION 3.3

Represent the reaction in part (c) of Question 3.2 by means of a reaction-coordinate diagram. On your diagram, label the following:

(a) the positions of the activated complexes;

(b) the energy that determines the rate of the reaction;

(c) the energy that determines the extent of the reaction at equilibrium.

Now that we have developed the basic principles of the mechanism of electrophilic aromatic substitution reactions, we shall consider five examples of this reaction that are of primary importance in synthetic organic chemistry. They all follow the general mechanism (Equation 3.7), but with a different X^+, so the first important task is to identify the nature of this electrophile in each case. The structure of the product will usually give a clue.

3.2 Nitration

The electrophilic aromatic substitution reaction that has received by far the most-detailed study is **nitration**, and consequently this is the one that provides us with the most-detailed mechanistic picture. Experimentally, nitration can be carried out with a mixture of concentrated nitric and sulfuric acids, the so-called 'nitrating mixture'. The product of the reaction with benzene is nitrobenzene, a yellow oily liquid with a sweet odour:

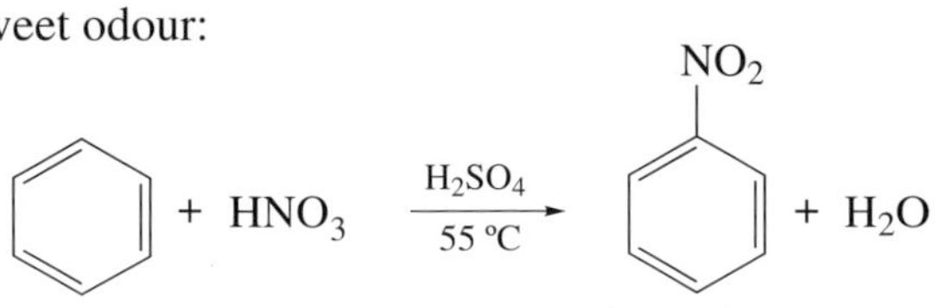

(3.8)

nitrobenzene

Let us look at the mechanism of this reaction. The principal problem in trying to formulate a mechanism for any electrophilic aromatic substitution reaction is to define the species that actually reacts with the benzene ring. The essential criterion is that it must be an electrophile, and it is usually a cation.

● Suggest a possible electrophile for the nitration reaction. (Look at the structure of the product, nitrobenzene.)

● As the NO_2, group has replaced H, it is reasonable to suggest that the electrophile is $\overset{+}{N}O_2$.

Indeed, the $\overset{+}{N}O_2$ ion, called the **nitronium ion**, is present in the nitrating mixture. Sulfuric acid is a stronger acid than nitric acid, and hence it can protonate HNO_3 to give $H_2\overset{+}{N}O_3$, which in turn can lose water, to form $\overset{+}{N}O_2$:

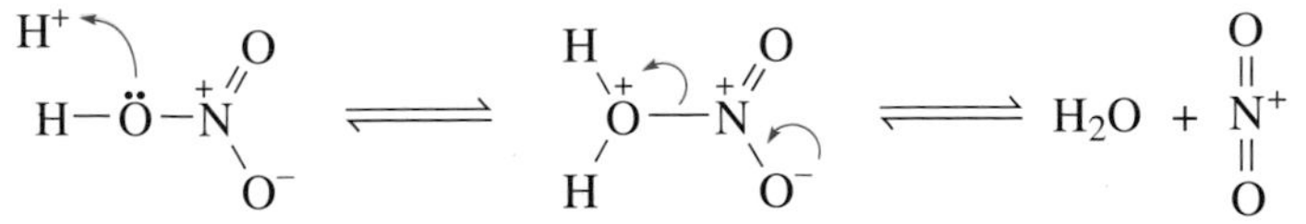

(3.9)

The existence of the nitronium ion is supported by the fact that stable nitronium salts have been prepared and isolated. Salts such as nitronium tetrafluoroborate ($\overset{+}{N}O_2\,BF_4^-$) and nitronium perchlorate ($\overset{+}{N}O_2\,ClO_4^-$) are powerful reagents for the nitration of aromatic compounds.

So, having identified the electrophile that in this case reacts with benzene to form the carbocation intermediate, it is a simple job to write down the mechanism, as shown in Scheme 3.1.

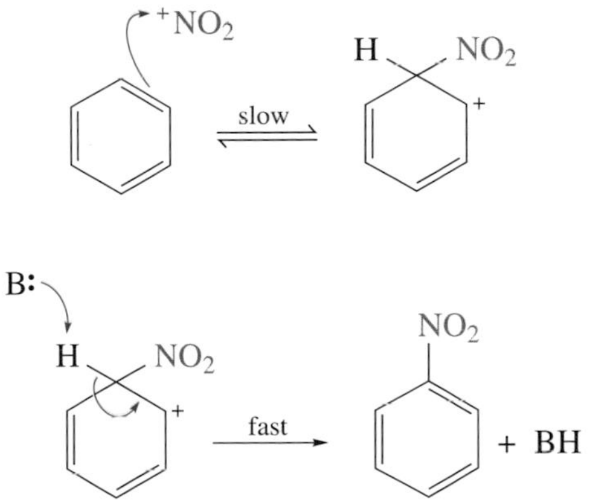

SCHEME 3.1

In this scheme, B represents a base, which could be H_2O, HSO_4^- or NO_3^-.

BOX 3.1 The dehydrating power of sulfuric acid

The formation of $\overset{+}{N}O_2$ from nitric acid arises by protonation and removal of water. Sulfuric acid is good at both of these jobs. Firstly it is a strong acid, and secondly it is a very good dehydrating agent. Much of the destructive power of sulfuric acid on organic tissue arises from dehydration. Hydrochloric acid is a similarly strong acid, but is much less corrosive because it does not absorb water as readily. When sulfuric acid absorbs water, much heat is generated, making the acid even more corrosive. In fact, the dehydrating power of sulfuric acid was used in one of the first successful attempts to make ice.

The first absorption machine was developed by Edmond Carré in 1850, using water and sulfuric acid (his brother, Ferdinand Carré developed the first ammonia/water refrigeration machine in 1859). Placing water in a vacuum led to evaporation of the water, but soon the atmosphere above the water was saturated with water vapour. The inclusion of concentrated sulfuric acid in a beaker alongside the water meant that the build up of water vapour was prevented by its being absorbed in the sulfuric acid. This meant the water evaporated further. The resulting cooling action eventually led to the formation of ice. On the other hand, the sulfuric acid solution heated up!

3.3 Halogenation

In **halogenation** reactions the halogen, Cl_2 or Br_2, reacts with benzene in the presence of a catalyst.

● Suggest an electrophile for the chlorination of benzene.

● By analogy with other electrophiles, the attacking reagent is Cl^+. Though the chloride ion, Cl^-, will be more familiar to you, Cl^+ is a perfectly acceptable and readily available electrophilic species.

The Cl^+ ion is formed via the action of a catalyst composed of a metal chloride, such as $SnCl_4$, $FeCl_3$, $SbCl_5$ or $AlCl_3$. The essential requirement is that it can act as a **Lewis acid**; that is, it behaves as an acceptor for a non-bonded electron pair on a Cl^- ion. Hence, $AlCl_3$ will form $AlCl_4^-$ and Cl^+ when reacting with a chlorine molecule:

$$Cl-Cl \; AlCl_3 \rightleftharpoons Cl^+ \; + \; {}^-AlCl_4 \qquad\qquad (3.10)$$

A study of the kinetics of the chlorination of benzene reveals that one molecule each of benzene, the halogen and the Lewis acid are involved in steps up to and including the slow, rate-limiting step. We can show the mechanism based on the generalized reaction (Equation 3.7) as follows:

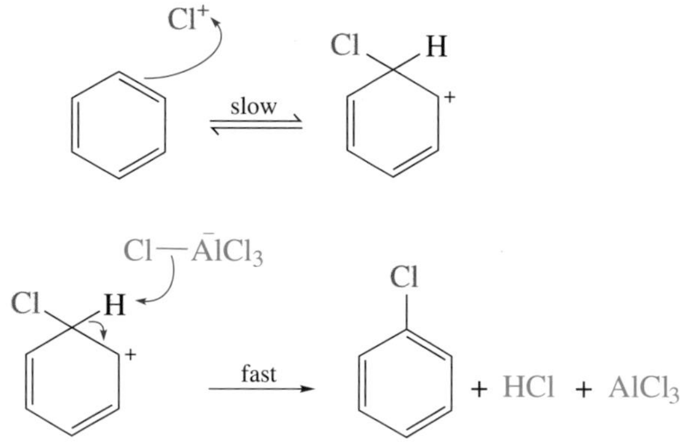

SCHEME 3.2

This is something of a simplified picture. In fact, the nature of the electrophile is not obvious. It may well be that no free Cl^+ ions are formed at all, but that the aluminium chloride (or any Lewis acid) serves only to polarize the halogen molecule; the actual electrophilic reagent would then be the positive end of the polarized chlorine molecule, rather than a free Cl^+ ion:

$$Cl_2 \; + \; AlCl_3 \rightleftharpoons \overset{\delta+}{Cl}-\overset{\delta-}{Cl}---AlCl_3$$

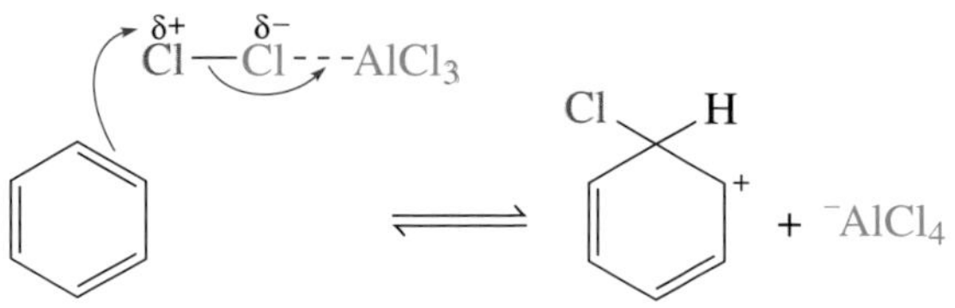

SCHEME 3.3

However, the chloride anion is certainly removed by the $AlCl_3$, as $AlCl_4^-$, in the rate-limiting step. For simplicity, we normally write the mechanism of halogenation using X^+ as the electrophile.

In the addition of a halogen molecule to an alkene (Part 1, Section 1.3) we considered that the electrons of the double bond polarized the halogen molecule sufficiently for reaction to occur. It is not surprising then that attack on the less-reactive benzene molecule requires additional polarization by a Lewis acid. (Indeed, more-reactive aromatic compounds react with halogens in the absence of any added Lewis acid.)

● Which halogens other than chlorine might be expected to react with benzene via electrophilic aromatic substitution?

○ Fluorine, bromine and iodine could be expected to undergo similar reactions with benzene to those that we have described for chlorine.

However, it is not quite that simple. Bromination takes place in much the same way as chlorination, with an appropriate catalyst, such as $FeBr_3$. However, although fluorine will react with benzene, the product is not fluorobenzene; rather, carbon–carbon bond rupture occurs, and other products are formed. Iodination fails because iodine is much less reactive than the other halogens. We shall discover ways of preparing both fluoro- and iodo-derivatives of benzene when we discuss some of the simple chemistry of diazonium salts in Section 5.

3.4 Sulfonation

Benzene will react with concentrated sulfuric acid at about 150 °C to produce benzenesulfonic acid*:

$$\text{benzene} + \text{conc. } H_2SO_4 \longrightarrow \text{benzenesulfonic acid (SO}_3\text{H)} + H_2O \qquad (3.11)$$

The mechanism involves the initial formation of sulfur trioxide (SO_3) from sulfuric acid. Though sulfur trioxide is a neutral molecule rather than a cation, it is still a relatively powerful electrophile.

* Although concentrated sulfuric acid is also part of the nitrating mixture (see Section 3.2), the conditions employed in the nitration reaction ensure that nitration rather than sulfonation is the predominant process.

The generation of sulfur trioxide from sulfuric acid, and the entire reaction sequence, is thought to occur as follows:

$$2H_2SO_4 \rightleftharpoons SO_3 + H_3O^+ + HSO_4^-$$

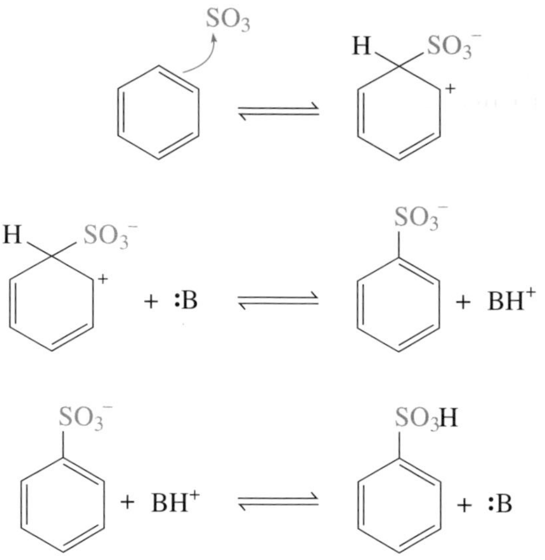

SCHEME 3.4

In this scheme, B represents a base, which could be H_2O or HSO_4^-.

● What do you notice as the main difference between this and the other electrophilic aromatic substitution reaction mechanisms we have discussed?

● All the four steps are depicted as equilibria.

Indeed, the overall **sulfonation** reaction is more correctly depicted as an equilibrium:

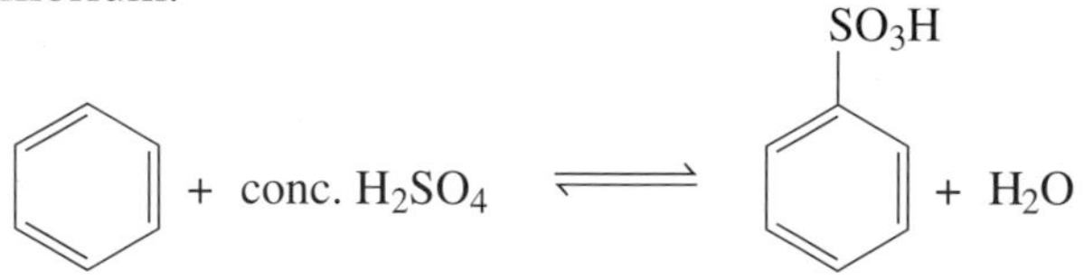

$$\text{benzene} + \text{conc. } H_2SO_4 \rightleftharpoons \text{benzenesulfonic acid} + H_2O \tag{3.12}$$

When the sulfur trioxide electrophile is generated with concentrated sulfuric acid, the position of the equilibrium favours benzenesulfonic acid. However, if we introduce more water — that is, use dilute sulfuric acid — we can convert benzenesulfonic acid back into benzene (a process called 'desulfonation'):

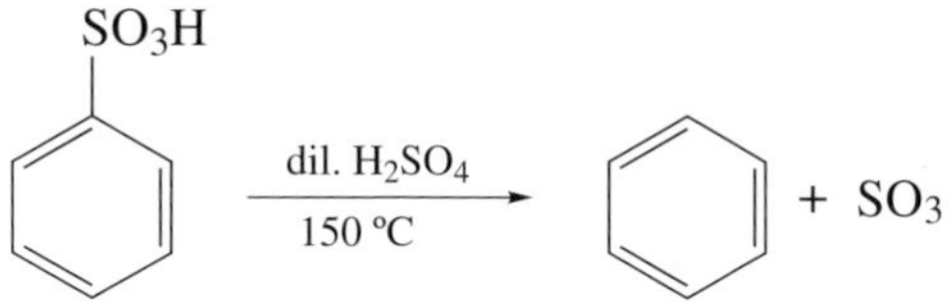

$$\text{benzenesulfonic acid} \xrightarrow[150\,°C]{\text{dil. } H_2SO_4} \text{benzene} + SO_3 \tag{3.13}$$

This is also an electrophilic aromatic substitution reaction, and the mechanism is exactly the same as in Scheme 3.4, except that the steps occur in the reverse order (in this case, H^+ is now the electrophile).

As a practical matter, aromatic sulfonic acids are not usually prepared and handled as such, because they tend to be very corrosive. They are usually prepared as intermediates in the synthesis of other compounds. One obvious comment, in view of the reversibility of sulfonation, is that the synthetic chemist can introduce a sulfonic acid group into a molecule at one stage in a synthesis, and remove it later as required. This tactic may seem somewhat pointless, but in fact it can be useful for protecting* a site in a molecule that would otherwise undergo an undesired reaction.

QUESTION 3.4

Represent the nitration of benzene by means of a reaction-coordinate diagram that reflects the mechanism we discussed in Section 3.2. (Don't worry about showing the formation of $\overset{+}{N}O_2$ in the diagram.)

QUESTION 3.5

Write a mechanism involving curly arrows for the desulfonation of benzenesulfonic acid.

QUESTION 3.6

What is the shape of the nitronium ion, $\overset{+}{N}O_2$?

3.5 Friedel–Crafts reactions

When chemists plan the synthesis of a complex organic molecule, one of the main concerns is building the correct carbon skeleton from smaller, more readily available compounds. To build the skeleton, carbon–carbon bonds need to be formed. The Friedel–Crafts reactions described in this Section represent one of the main methods of adding carbon chains to aromatic rings.

● Bearing in mind that electrophiles, X^+, are necessary for the formation of C(aromatic)—X bonds in electrophilic aromatic substitution reactions, what kind of reagent do you think will be necessary to make a C(aromatic)—C bond?

● The obvious type of electrophile to use to make C—C bonds in this case is a carbocation (**3.3**).

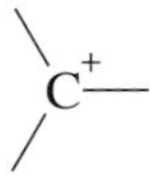

3.3

We shall examine two types of Friedel–Crafts reaction, alkylation and acylation, in which carbocations (or species that behave like carbocations) are generated by the reaction of a haloalkane (RCl) or an acyl chloride (RCOCl, **3.4**) with AlCl$_3$:

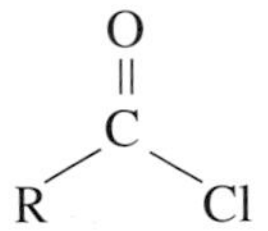

3.4

ALKYLATION 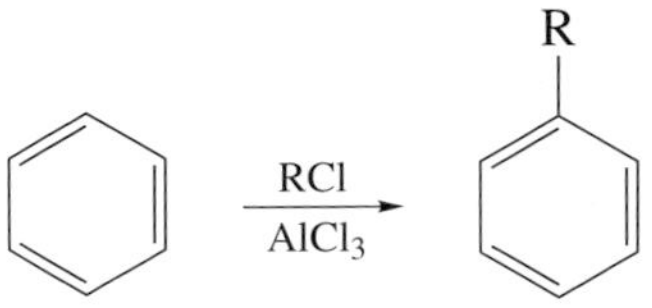 (3.14)

ACYLATION 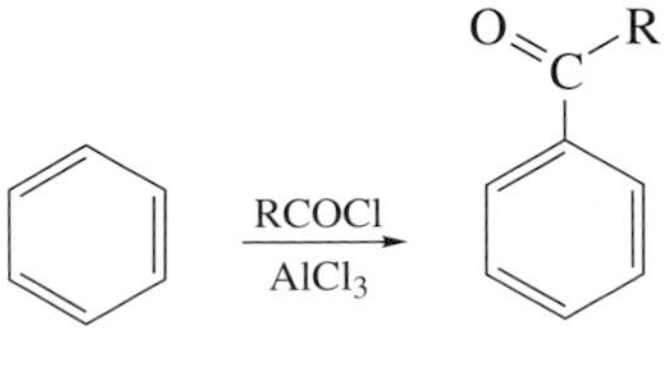 (3.15)

---

* Methods of protecting particular sites in molecules during a synthetic procedure are discussed in *Mechanism and Synthesis*[2].

BOX 3.2 Charles Friedel

Charles Friedel was born in Strasbourg on 12 March 1833. He studied both chemistry and mineralogy at the Sorbonne in Paris. In 1856 he was made curator of the collection of minerals at the École des Mines, and in 1871 he became an instructor at the École Normale. In 1876 he was appointed professor of mineralogy at the Sorbonne, and was Professor of Organic Chemistry there from 1884 to 1899. Friedel collaborated with Crafts during the period 1874–91, and in 1877 they discovered the Friedel–Crafts reaction. He synthesised a variety of new compounds, and distinguished the function of the alcohol and carboxyl functional groups in lactic acid. From 1879 to 1887 Friedel worked on the synthesis of minerals, including diamonds, using heat and pressure.

An extract from the notebook used by Friedel is shown in Figure 3.3.

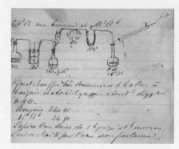

Figure 3.3
A page from a detailed notebook of Friedel's experimental work.

Figure 3.2
Charles Friedel 1832–1899.

BOX 3.3 James Mason Crafts

James Mason Crafts was born in Boston, Massachusetts on 8 March 1839. He graduated from Harvard in 1858, after which he spent seven years at the Freiberg mining school, the University in Heidelberg, and the École des Mines in Paris. On his return to the United States he became Professor of Chemistry at Cornell (1868–70), and then at the Massachusetts Institute of Technology (1870–1880). He also collaborated with Friedel on the synthesis of the esters of silicic acid. Besides several honours received in the US and abroad, he was made Chevalier of the Legion of Honneur by the French government in 1885.

Figure 3.4
James Mason Crafts 1839–1917.

3.5.1 Friedel–Crafts alkylation

In 1877, Charles Friedel and James Crafts reported that benzene and 2-chloropropane reacted together in the presence of $AlCl_3$ to form 2-phenylpropane. This type of reaction is now generally known as **Friedel–Crafts alkylation**.

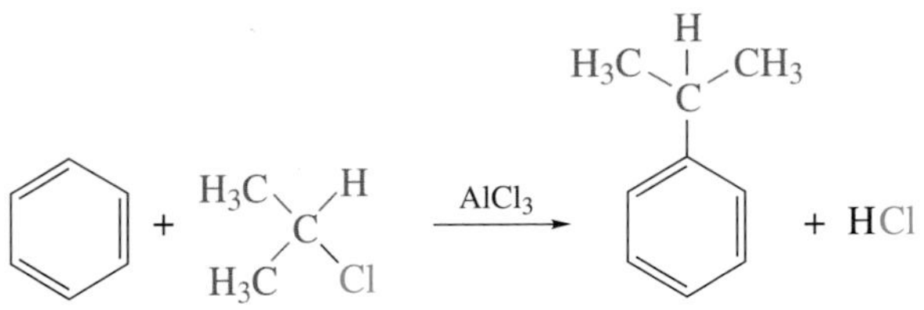

$$(3.16)$$

2-phenylpropane

● By analogy with chlorination, what do you think is the role of the $AlCl_3$?

● The $AlCl_3$ is acting as a Lewis acid, and is either helping to polarize the $C-Cl$ bond, in the sense $C^{\delta+}-Cl^{\delta-}$, or forming a true carbocation intermediate.

The equilibrium reaction between $AlCl_3$ and the haloalkane can be written as follows:

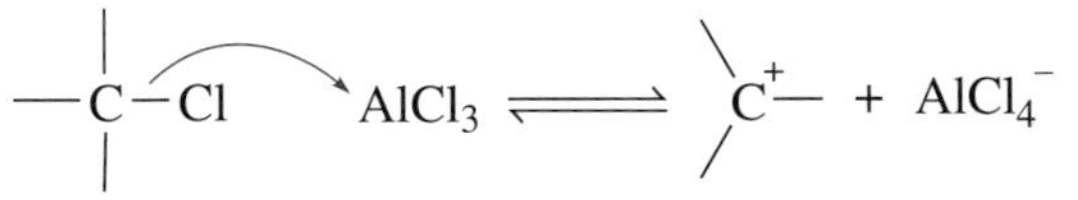

$$(3.17)$$

Although it is rather unlikely that a free carbocation is formed in alkylation reactions, for convenience we generally write the mechanism as if it were. So, for the original Friedel–Crafts alkylation, we can write the mechanism as:

$$(CH_3)_2CHCl + AlCl_3 \rightleftharpoons (CH_3)_2\overset{+}{C}H + AlCl_4^-$$

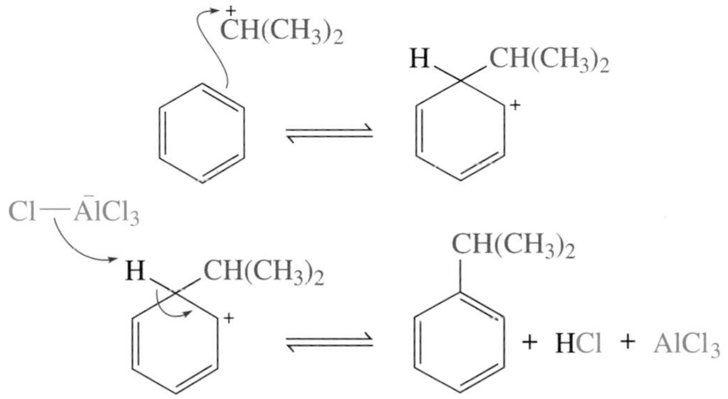

SCHEME 3.5

This is directly analogous to other electrophilic aromatic substitution reactions, and the reaction-coordinate diagram is also very similar. This is shown in Figure 3.5 (ignoring the formation of the electrophile). The reaction of the electrophile (the carbocation) with the benzene ring, is again the rate-limiting step.

Friedel–Crafts alkylations usually use a chloroalkane/$AlCl_3$ combination to generate the carbocation, but other methods are often equally successful, and other Lewis acids, such as $TiCl_4$ and $SbCl_5$, may be used. Today, many Friedel–Crafts reactions are carried out in industry using a zeolite[*] as the catalyst; this avoids the generation of hazardous acid waste and the need to dispose of metal halides. The manufacture of 2-phenylpropane (cumene) on an industrial scale uses propene as the starting material, and the carbocation is produced by addition of a proton:

* The structures and uses of zeolites are discussed in the Case Study with *Chemical Kinetics and Mechanism*[3].

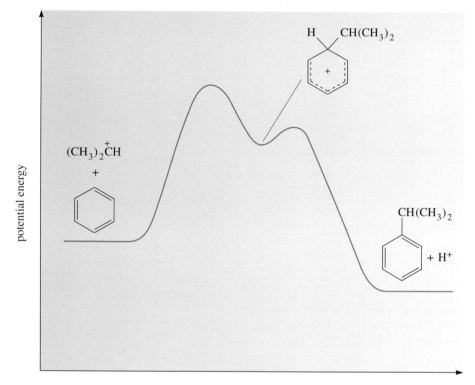

Figure 3.5
The reaction-coordinate diagram for the Friedel–Crafts reaction of benzene with 2-chloropropane in the presence of aluminium chloride.

$$CH_3-CH=CH_2 + H^+ \longrightarrow CH_3-\overset{+}{C}H-CH_3 \qquad (3.18)$$

As you will see in the Case Study at the end of this Book, cumene is used for the industrial preparation of phenol and propanone (acetone) in large amounts.

Despite the efficient synthesis of 2-phenylpropane, Friedel–Crafts alkylation is not generally a very useful route to aromatic compounds with alkyl side-chains. In 1878, less than one year after the first Friedel–Crafts synthesis, a chemist called G. Gustavson tried to make 1-phenylpropane from 1-chloropropane and benzene, using AlCl$_3$ as catalyst. To his surprise, every time he tried the reaction, the isolated product was the isomer, 2-phenylpropane:

$$(3.19)$$

It was shown subsequently that the reason for the formation of this initially unexpected product is that the intermediate carbocation **rearranges**.

● What is the order of stability of primary, secondary and tertiary carbocations?

● The order of stability is:

$$R_3\overset{+}{C} > R_2\overset{+}{C}H > R\overset{+}{C}H_2$$

tertiary secondary primary

● What sort of carbocation is produced from 1-chloropropane and $AlCl_3$?

● The least-stable carbocation, the primary carbocation $CH_3CH_2\overset{+}{C}H_2$.

It is a general observation that carbocations rearrange to form more-stable carbocations. In this case, the primary carbocation, $CH_3CH_2\overset{+}{C}H_2$, rearranges, formally by a hydride (H^-) shift, to form the secondary carbocation,

$$CH_3CH_2\overset{+}{C}H_2 \longrightarrow CH_3\overset{+}{C}HCH_3 \qquad \text{via } CH_3-\overset{\overset{H}{|}}{\underset{\underset{H}{|}}{C}}-CH_2^+ \qquad (3.20)$$

As well as hydride shifts, alkyl shifts are also possible; for example

$$\underset{H_3C}{\overset{H_3C}{\diagdown}}\overset{}{\underset{}{C}}-CH_2Cl \xrightarrow{AlCl_3} \underset{H_3C}{\overset{H_3C}{\diagdown}}\overset{}{\underset{}{C}}-\overset{+}{C}H_2 \rightleftharpoons \underset{H_3C}{\overset{H_3C}{\diagdown}}\overset{+}{\underset{}{C}}-CH_2CH_3 \qquad (3.21)$$

primary   tertiary

● What group has migrated in this case?

A $-CH_3$ group.

It is frequently difficult to predict the way in which a particular carbocation is going to rearrange, but more-stable carbocations are always formed preferentially. Secondary carbocations will form tertiary carbocations on rearrangement, but not primary ones. So, one problem with Friedel–Crafts alkylation reactions is that unless you start with a chloroalkane that gives rise to the most-stable carbocation possible, you will end up with a mixture of products, or even none of the desired product.

● Which of the following chloroalkanes will give an unrearranged product on reaction with benzene in the presence of $AlCl_3$?

CH_3CH_2Cl, $(CH_3)_3CCl$, $CH_3CH_2CH_2CH_2Cl$, $C_6H_5CH_2CH_2Cl$ and $(CH_3)_2CHCH_2Cl$

● Only the first two chloroalkanes will give an unrearranged product: $CH_3CH_2^+$, cannot rearrange to anything other than a primary carbocation, and $(CH_3)_3C^+$ is already a tertiary carbocation. All the others give rise to primary carbocations that are able to rearrange to more-stable carbocations.

There is one more problem with Friedel–Crafts alkylations which limits their use in synthesis. In Section 4, you will learn about the influence that substituents on the benzene ring have on the rate of electrophilic aromatic substitution. Alkyl groups **activate** the ring to further substitution; that is, alkylbenzenes undergo aromatic substitution faster than benzene itself.

● What effect will this have on the Friedel–Crafts alkylation of, say, benzene?

• If the product is more reactive than the starting material — benzene in this case — then the final mixture will contain products in which more than one hydrogen in the ring has been substituted by an alkyl group.

This reaction, where the product contains several compounds with more than one alkyl group in the ring, is known as **polyalkylation**.

Many of the problems associated with Friedel–Crafts alkylations are solved by the use of acylation instead of alkylation, as described in the next Section.

3.5.2 Friedel–Crafts acylation

Friedel–Crafts acylations are again quite typical electrophilic aromatic substitution reactions, in which the electrophile is an **acylonium ion (3.5)** or equivalent. The generation of acylonium ions is straightforward, and involves a similar equilibrium to that in the alkylation reactions:

$$R-\overset{+}{C}=O \longleftrightarrow R-C\equiv\overset{+}{O}$$

3.5

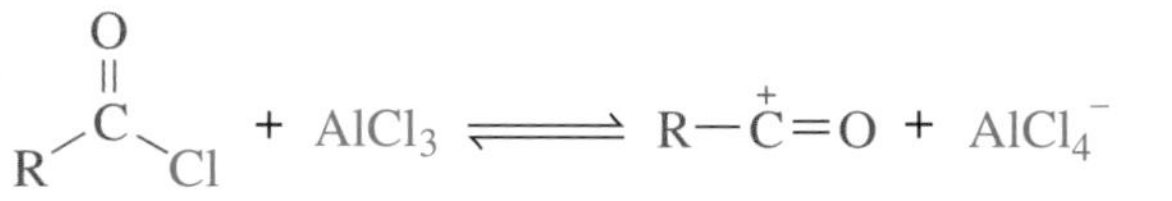

$$\underset{R}{\overset{O}{\overset{||}{C}}}\diagdown_{Cl} + AlCl_3 \rightleftharpoons R-\overset{+}{C}=O + AlCl_4^- \tag{3.22}$$

The advantages of using acylation in synthesis are that (i) acylonium ions do not rearrange, and (ii) the acyl group **deactivates** the ring to further substitution; that is, the ketone product, C_6H_5COR, is *less* reactive to attack by $R\overset{+}{C}O$ than benzene itself.

The source of acylonium ions is also not limited to acid chlorides, since anhydrides (**3.6**), also generate acylonium ions with $AlCl_3$; for example

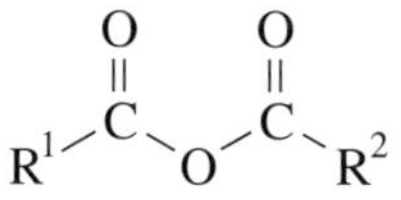

$$R^1\diagdown\underset{O}{\overset{O}{\overset{||}{C}}}\diagdown_O\diagup\underset{}{\overset{O}{\overset{||}{C}}}\diagup R^2$$

3.6

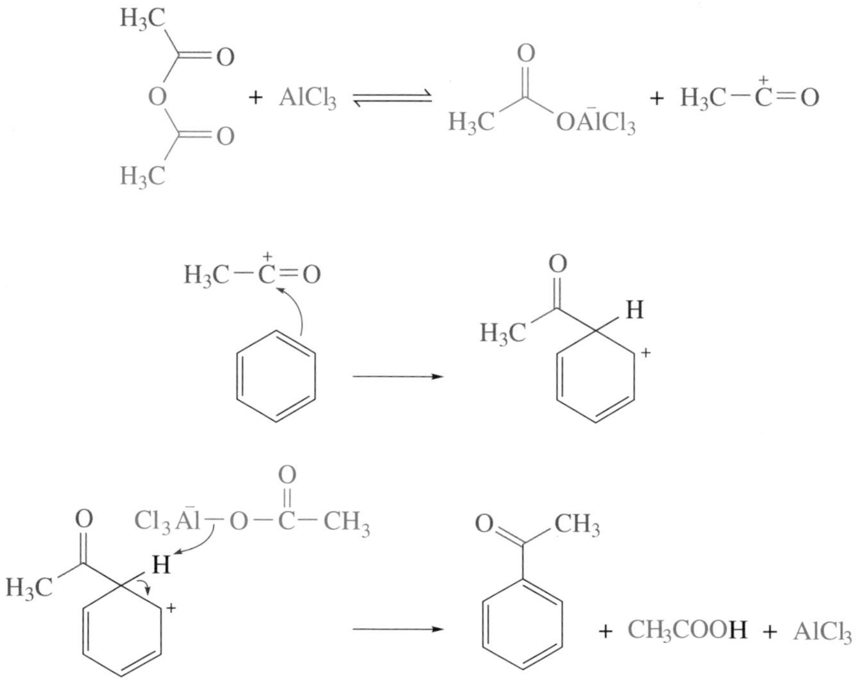

SCHEME 3.6

The acyl-substituted aromatic compound is a very versatile synthetic intermediate. The corresponding monoalkylated benzenes, formally derived from the primary carbocation RCH_2^+, can easily be prepared in high yield by the **Clemmensen reduction,** which involves reaction of the acyl compound with zinc amalgam and concentrated hydrochloric acid:

$$(3.23)$$

3.6 Summary of Sections 3.2–3.5

1 The principal problem in formulating the mechanism of an electrophilic aromatic substitution reaction is identifying the electrophile that reacts with the benzene ring. The structure of the product will usually give a clue.

2 *Nitration*   The electrophile is $\overset{+}{N}O_2$, the nitronium ion, usually generated by the so-called nitrating mixture of concentated HNO_3/H_2SO_4. In this nitrating mixture, H_2SO_4 is the stronger acid, and so HNO_3 acts as a base.

3 *Bromination and chlorination*   The electrophiles are best described as Br^+ and Cl^+, respectively. However, a Lewis acid catalyst is usually needed. There is evidence to show that the halogen molecule is polarized by the catalyst, and that these two molecules together provide the effective electrophile:

$$Cl_2 + AlCl_3 \rightleftharpoons \overset{\delta+}{Cl}-\overset{\delta-}{Cl}\text{-}\text{-}\text{-}AlCl_3 \qquad (3.24)$$

4 *Sulfonation*   The reaction takes place only in concentrated sulfuric acid, and the electrophile is not a cation, but sulfur trioxide (SO_3). Sulfonation is reversible.

5 *Alkylation*   In general, reagents that can ionize to form a carbocation can alkylate benzene, and the electrophile can be written simply as shown in Structure **3.3**. As in halogenation, Lewis acid catalysts are of importance for Friedel–Crafts reactions with haloalkanes and alkenes; in these cases the electrophile can be considered to be the carbocationic species R^+, formed as follows:

$$RCl + AlCl_3 \rightleftharpoons R^+ + AlCl_4^- \qquad (3.25)$$

The electrophilic carbocation can undergo rearrangement to give a mixture of products with benzene. Rearrangement of primary carbocations to secondary carbocations means that Friedel–Crafts alkyation is not suitable for the synthesis of primary alkyl benzene derivatives. There is also the possibility of polyalkylation, in which more than one alkyl group is introduced into the benzene ring.

6 *Acylation*   This substitution reaction involves the acylonium ion, $R\overset{+}{C}O$, as the electrophile:

$$(3.22)$$

3.3
alkylation
electrophile

Acylation does not suffer from the problems that alkylation does (see point 5). In practice, therefore, benzene is often first acylated with an acid chloride, say, and then the Clemmensen reduction is used to reduce the acyl substituent to a primary alkyl group in high yield.

QUESTION 3.7

Write mechanisms for the following reactions, and, where appropriate, explain the distribution of the products. Identify the electrophile that reacts with the benzene nucleus in each case.

(a) benzene + $CH_3CH_2CH_2Cl$ $\xrightarrow{AlCl_3}$ (benzene with $CH_2CH_2CH_3$) 30% + (benzene with $H_3C-\underset{H}{\overset{|}{C}}-CH_3$) 70%

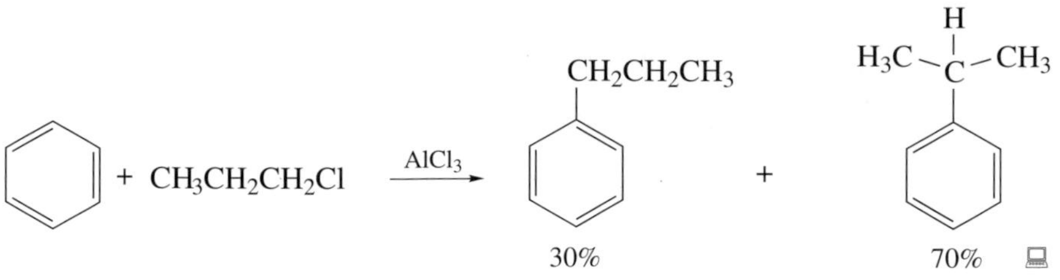

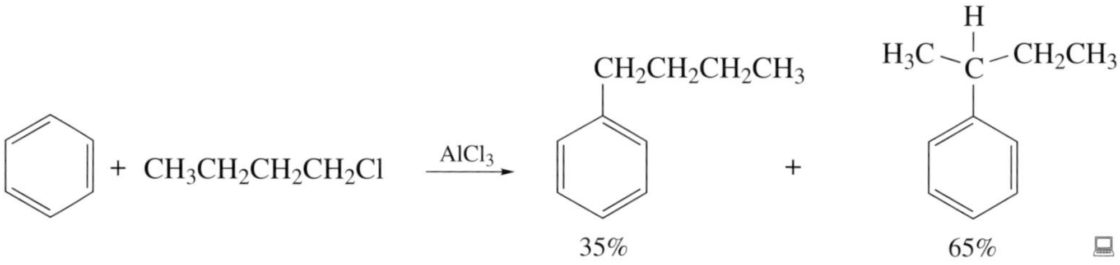

(b) benzene + $CH_3CH_2CH_2CH_2Cl$ $\xrightarrow{AlCl_3}$ (benzene with $CH_2CH_2CH_2CH_3$) 35% + (benzene with $H_3C-\underset{H}{\overset{|}{C}}-CH_2CH_3$) 65%

(c) benzene + (ethanoic anhydride: CH_3CH_2 and CH_3CH_2 with O, O) $\xrightarrow{AlCl_3}$ (benzene with $O=\!\!\!<\!\!CH_2CH_3$) + CH_3CH_2COOH

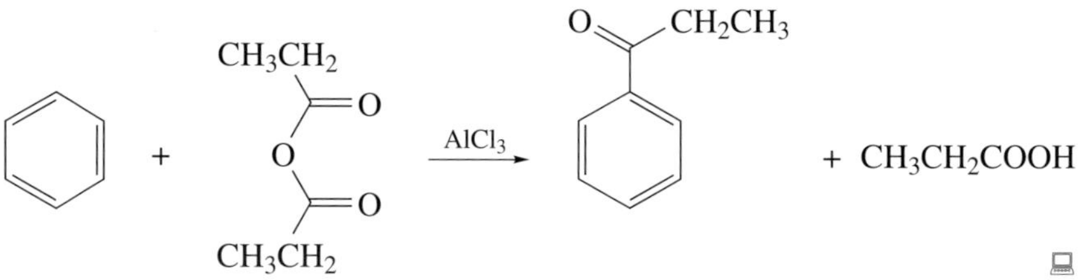

THE EFFECTS OF SUBSTITUENTS

Once one substituent has been introduced into the benzene ring, there are three unique positions where a second electrophilic substitution can take place. For example, there are three positional isomers of methoxymethylbenzene:

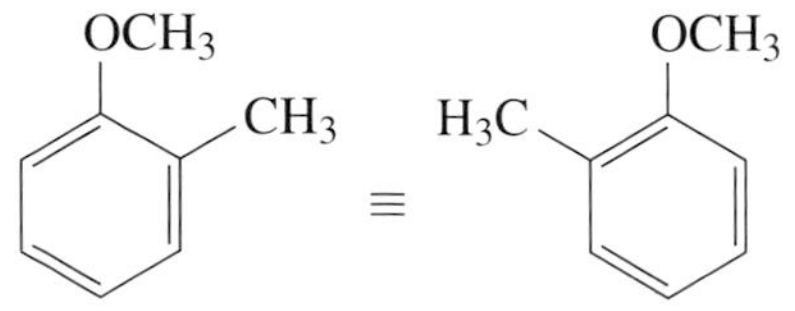

ortho-methoxymethylbenzene **4.1** 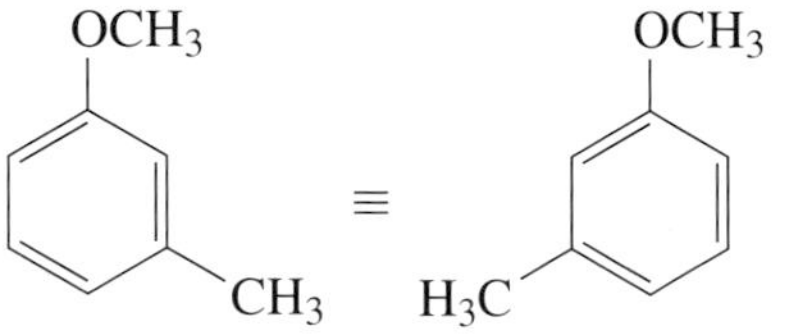   *meta*-methoxymethylbenzene **4.2** 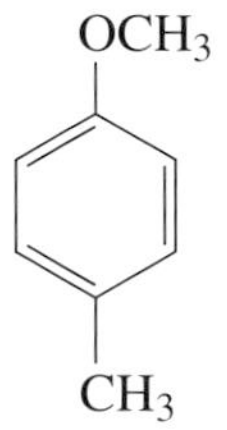   *para*-methoxymethylbenzene **4.3**

So if we consider the synthesis of any disubstituted benzene derivative from a monosubstituted derivative we might expect that three products would be formed in statistical amounts: we might expect to obtain two parts *ortho*, two parts *meta* and one part *para* substitution:

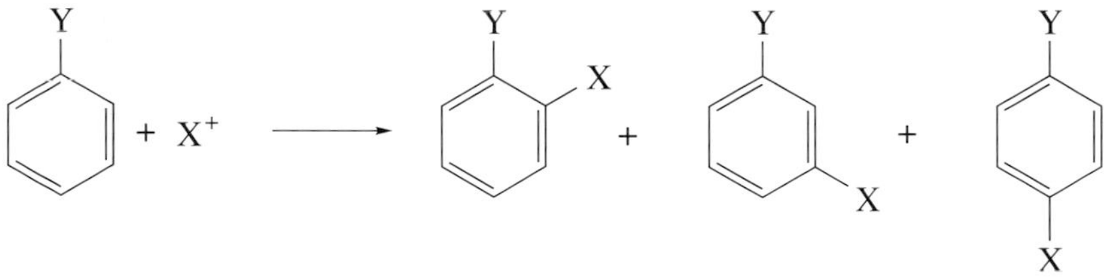

(4.1)

40%   40%   20%

statistically expected disubstitution pattern

The proportions of products that are usually found in such reactions do not approach this statistical distribution. This is illustrated by the nitration of chlorobenzene and the chlorination of nitrobenzene. The distributions of products for these reactions are shown in Figure 4.1, from which you can see that the isomer distribution varies according to which substituent is originally present in the monosubstituted benzene.

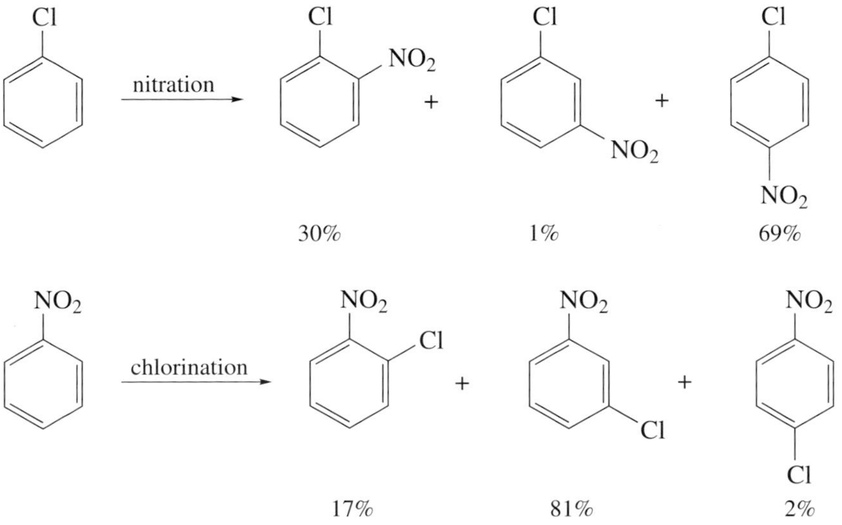

30%   1%   69%

17%   81%   2%

Figure 4.1
The distribution of products formed in the nitration of chlorobenzene, where the *ortho* and *para* isomers predominate, and in the chlorination of nitrobenzene, where the *meta* isomer predominates.

What is more, it destabilizes the carbocations that result from *ortho/para* attack more effectively than those that result from *meta* attack. Thus, of the possible evils, *meta* substitution is the least evil! Let's look at this in more detail.

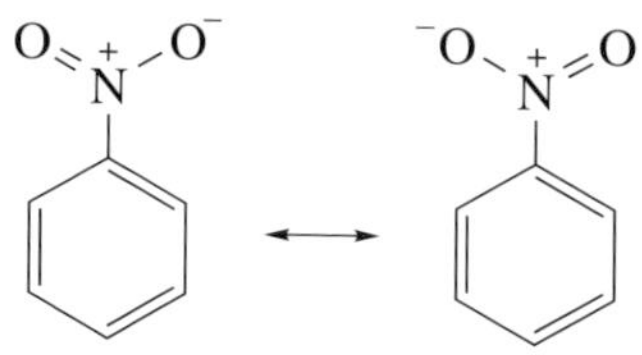

The nitro group in nitrobenzene has two resonance forms (**4.9**), both of which carry a positive charge on the nitrogen atom.

4.9

Attack of an electrophile such as Cl^+ at the *ortho* or *para* positions of nitrobenzene results in unfavourable interaction between the positive charge of the carbocation and the positive charge on the nitrogen of the nitro group. We can show this by drawing out the resonance structures for chlorination at the *para* position (**4.10**) as an example:

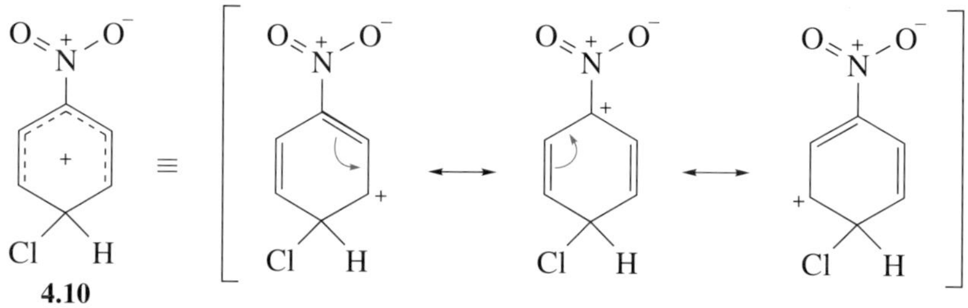

4.10

Thus, even though one of the resonance structures is a tertiary carbocation, it is destabilized by the mutual repulsion of the two positive charges on adjacent atoms. A similar situation occurs for *ortho* attack, but not for *meta* attack (**4.11**). Hence *meta* substitution is preferred:

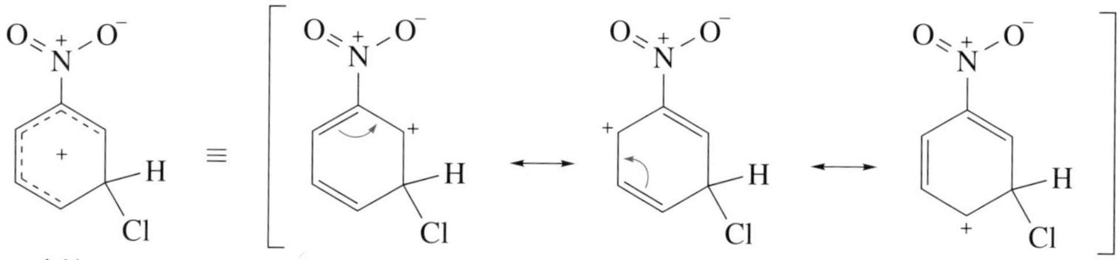

4.11

You will not be surprised to discover that any other groups that can provide a similar, unstable charge distribution, such as $-\overset{+}{N}H_3$, result in the *meta* isomer predominating. (Note that the $-NH_2$ group, which is *ortho/para*-directing, is converted almost entirely to the *meta*-directing $-\overset{+}{N}H_3$ group in strong acid solution.) Carbonyl (**4.12**) and cyano (**4.13**) groups are also *meta*-directing by virtue of the fact that these are polarized such that a partial positive charge resides on the carbon atom of these two groups.

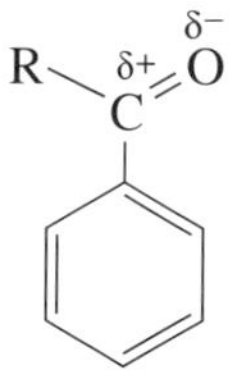

4.12

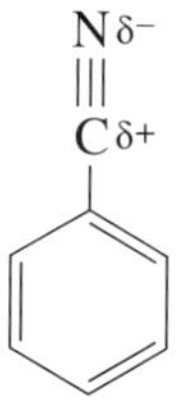

4.13

Table 4.1 summarizes the directing effects of a variety of functional groups. This information is reproduced in the *Data Book* (available from the CD-ROM).

Table 4.1 The directing effects of various substituents attached to the benzene ring

ortho/para-directing groups	*meta*-directing groups
R	NO_2
OH, OR	$\overset{+}{N}H_3$, $\overset{+}{N}R_3$
NH_2, NHR, NR_2	SO_3H, SO_3R
NHCOR	CHO, COR
Cl, Br, I	CN

Although the statistically expected pattern for *ortho* : *para* product distribution is 2 : 1, in practice (as we have seen in the nitration example in Figure 4.1) the proportion of the *para* product obtained is rather greater than 33%. This is explicable in terms of the greater steric hindrance to attack at the *ortho* positions.

BOX 4.1 Trinitrotoluene, TNT

TNT stands for trinitrotoluene or, more correctly 2,4,6-trinitrotoluene. It is formed by nitration of toluene, and, since the methyl group is *ortho/para* directing, the nitro groups go into these three positions. Whichever position the nitro group goes into first, the positions *ortho/para* to the methyl group are always *meta* to this nitro group:

$$(4.3)$$

However, because the nitro group is deactivating, attaching the last nitro group to dinitrotoluene by an electrophilic aromatic substitution requires high temperatures!

TNT forms pale yellow crystals with a melting temperature of 82 °C, which means it can be melted and poured into artillery shells and other explosive devices. In the absence of a detonator, it is quite a stable material. However, when detonated, it quickly changes from a solid into hot expanding gases — hence the explosive force. Two moles of solid TNT almost instantly change to 15 moles of hot gases plus some powdered carbon, which gives a dark sooty appearance to the explosion!

$$2C_7H_5N_3O_6(s) = 3N_2(g) + 7CO(g) + 5H_2O(g) + 7C(s) \qquad (4.4)$$

In the trenches of the First World War, shells containing TNT were therefore known as coalboxes.

Now see if you have understood the directing effects of substituents by tackling the following Questions.

QUESTION 4.1

(a) What are the principal products of the reaction of methoxybenzene with: (i) H_2SO_4; (ii) HNO_3/H_2SO_4; (iii) $Cl_2/FeCl_3$?

(b) What are the products of the reaction of nitrobenzene with: (i) H_2SO_4; (ii) HNO_3/H_2SO_4; (iii) $Cl_2/FeCl_3$?

5

DIAZONIUM SALTS

5.1 From coal tar to dyes

In Section 1 we speculated that the beginning of the organic chemical industry in the middle of the nineteenth century was based on coal tar and the numerous compounds contained within it, such as benzene. A benzene derivative that was of great interest to W. H. Perkin (Box 5.1, p. 70) — while working at the Royal College of Chemistry in 1856 — was aminobenzene or aniline (**5.1**).

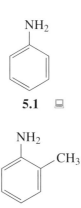

5.1

While engaged on an unsuccessful synthesis of quinine (Figure 5.1) from *ortho*-methylaniline (**5.2**), Perkin tried the same reaction with aniline. He obtained a black precipitate, which, on further examination, he found to contain the now-famous dye called 'mauve' (now known as 'mauveine') or 'aniline purple' (Figure 5.2). This discovery led to the first dye to be synthesised commercially, and it could not have arrived at a better time. Textile production was booming, and hence there was an increased demand for dyestuffs, all of which had previously been obtained from natural sources, usually plants. Perkin's mauve dye created considerable excitement in the United Kingdom and soon became popular in France also. Queen Victoria wore a mauve dress at the London International Exhibition of 1862 (Figure 5.3), penny postage stamps were printed with mauve (see front cover of this book) and, according to the magazine *Punch*, the London policemen directed loiterers to 'get a mauve on'. Dyestuffs manufacture became a significant industry, not only in Britain but also in Germany, and later led to the development of pharmaceuticals (Box 5.2, p. 71).

5.2

Figure 5.1
Quinine is a vital constituent of Indian Tonic water. It was used in India because of its antimalarial properties.

Figure 5.2
The original mauve dye synthesised by Perkin.

Figure 5.3
A dress dyed with mauveine as worn by Queen Victoria at the London International Exhibition.

Interestingly enough, Perkin was lucky when he first prepared mauveine. It is impossible to make this dye if only aniline is present, and in fact his samples of aniline almost certainly contained two isomers of methylaniline as well. We have shown the formation of mauveine in Figure 5.4 in a schematic way that gives the right answer but does not give any indication of the mechanism. However, you can see from this that we have the *ortho* and *para* isomers of methylaniline as reactants as well as aniline.

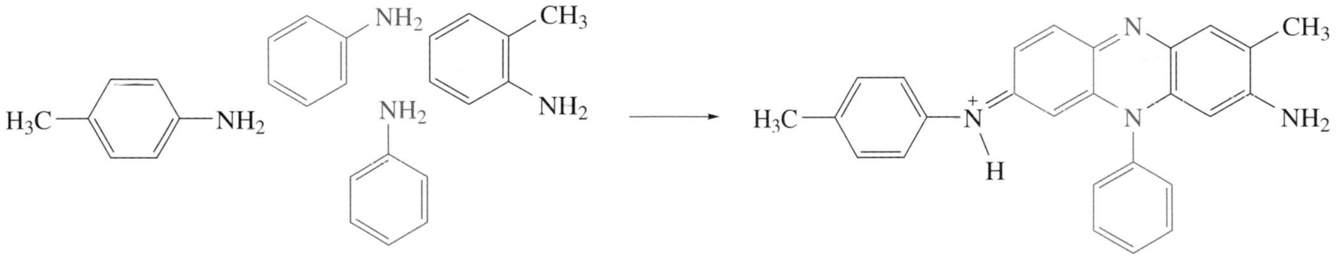

Figure 5.4 A schematic diagram to show the formation of mauveine from aniline and the *ortho* and *para* isomers of methylaniline.

The demand for mauve meant a demand for aniline. Aniline had previously been prepared by the distillation of a natural blue dyestuff obtained from the indigo plant, but in this age of coal tar it was soon replaced by a synthesis from benzene. Benzene is first nitrated, and the product reduced by hydrogen generated from iron or tin and dilute hydrochloric acid, to yield aniline, which can be purified by distillation (Scheme 5.1).

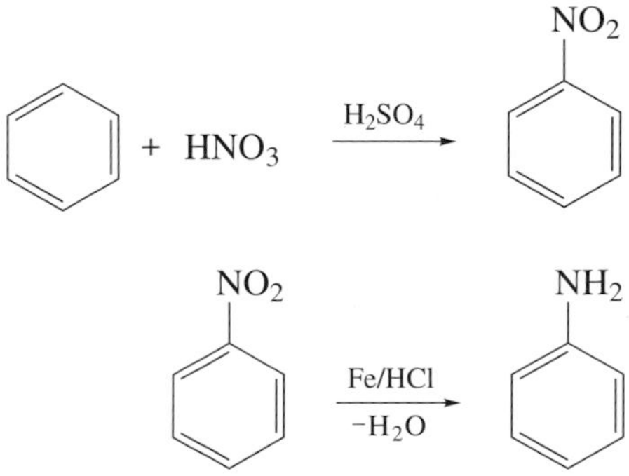

SCHEME 5.1

BOX 5.1 William Henry Perkin

William Henry Perkin was born on 12 March 1838 in London. Perkin's father, who was a builder by trade, wanted him to become an architect, but, thanks to his schoolteacher, Thomas Hall, he was excited by chemistry. At the age of 15, he studied with A. W. Hofmann (of elimination fame) at the Royal College of Chemistry (now Imperial College), and became Hofmann's assistant at the age of 17. In 1856, during the Easter vacation, Perkin, aged 18, synthesised mauveine in his homemade laboratory. He patented his method and left the Royal College of Chemistry (much to Hofmann's annoyance). With his father's support, he then commercialized his product. In 1857 his father and brother joined him in his works on the banks of the Grand Union Canal at Greenford (Figure 5.6). In 1874, Perkin sold his works to Brook Simpson and Spiller (Spillers is now known as British Bakeries), but continued his research in his laboratory at home. The mauve dye was one of the first truly industrial chemicals, and this was followed by Britannia violet and Perkin's green (the water of the nearby canal was said to change colour every week according to which dye was being manufactured!). From 1884, he conducted research into magnetic-induced rotation of plane-polarized light, using the results to clarify the structures of organic molecules.

Figure 5.5
William Henry Perkin 1838–1907.

Figure 5.6
Perkin's dye works in Greenford.

Another group of coal tar dyes that should be mentioned here are the so-called azo dyes. In preparing azo dyes, the azo group ($-N=N-$) is always obtained from a diazonium salt, which in turn is prepared from an aromatic amine. The simplest is a benzenediazonium salt, which itself is derived from aniline:

$$\text{(structure: aniline } \xrightarrow[\text{NaNO}_2]{0\,^\circ\text{C/HCl}} \text{ benzenediazonium chloride)}$$

benzenediazonium chloride

(5.1)

BOX 5.2 From dyes to drugs

Despite Perkin's success, the development of the British synthetic dyestuffs industry was very slow. In other countries, chemists were quick to see the potential and many new synthetic dyes were made from the myriad chemicals in coal tar, with Germany dominating the industry. Like many 'natural' dyes, the yellow alizarin (**5.3**) and indigo are simple molecules, but until 1868 their chemical structures were completely unknown. In that year, alizarin was shown chemically to be derived from the hydrocarbon anthracene, obtained from coal tar. In June 1869, Perkin and the German dye company BASF filed patents for the same synthetic route to alizarin just one day apart. As a result, the industry producing the natural dye was killed off almost overnight. Perkin's company alone was producing over 400 tonnes of synthetic alizarin a year, and at less than half the price of the natural product.

However, the synthesis of dyes led to the modern pharmaceutical industry. In his search for antiseptic substances that could be used to treat bacterial infections, a German physician, Paul Ehrlich, discovered that one of the dyes he used for staining his microscope slides actually killed bacteria. He found that the yellow dye flavine killed the germs responsible for abscesses. This led to the development of a wide range of useful drugs all derived from chemicals found in coal tar.

5.3

The mechanism involves the formation of nitrous acid from hydrochloric acid and sodium nitrite. Nitrous acid is fairly unstable and so has to be formed below 5 °C just before it is used:

$$HCl + NaNO_2 \longrightarrow HNO_2 + NaCl \tag{5.2}$$

nitrous acid

In acidic solution the nitrous acid becomes protonated, and loses water to form a nitrosonium cation, $\overset{+}{N}=O$:

$$H-O-N=O \underset{}{\overset{H^+}{\rightleftharpoons}} \quad \overset{+}{O}-N=O \rightleftharpoons \overset{+}{N}=O + H_2O \tag{5.3}$$

Like the nitronium ion, the nitrosonium ion is electrophilic, and so will react with a nucleophilic amino group:

$$\text{Ph}-\overset{..}{N}H_2 \quad \overset{+}{N}=O \longrightarrow \text{Ph}-\overset{+}{N}H_2-N=O \tag{5.4}$$

This species then loses a proton and rearranges to form a nitrogen–nitrogen double bond:

$$\overset{+}{N}H-N=O \quad \overset{-H^+}{\longrightarrow} \quad -NH-N=O \longrightarrow -N=N-OH \tag{5.5}$$

The final step involves protonation of the oxygen and loss of water to generate the diazonium cation:

$$-N=N-OH \quad \overset{H^+}{\longrightarrow} \quad -N=N-\overset{+}{O}H_2 \longrightarrow -\overset{+}{N}\equiv N \tag{5.6}$$

diazonium cation

Diazotization, as the reaction is called, is generally carried out in the following way. The amine is added to an aqueous solution of hydrochloric acid. This mixture is then cooled to around 0 °C, and a solution of sodium nitrite is added at such a rate that the temperature does not rise above 5 °C (the reaction is exothermic). Because diazonium salts slowly decompose, even at ice-bath temperatures, the solution is used immediately after preparation. The large number of reactions undergone by diazonium salts may be divided into two classes: substitution reactions, in which nitrogen is lost as N_2, and some other atom or group is attached to the benzene ring in its place; and coupling reactions, in which nitrogen remains in the product. The latter provide a means of preparing azo dyes, first discovered by the German chemist Peter Griess (Box 5.3).

BOX 5.3 Peter Griess

The first azo dyes were discovered by Peter Griess in 1864. Griess was born in Kirchhosbach in Germany in 1829 of middle-class Prussian parents. At the age of 16, he went to agricultural school but lost interest in agriculture, and went on to Jena and Marburg Universities, where he drifted between disciplines. At this point in his life, he became something of a reprobate, ending up in prison and in severe financial debt. To pay off his debts, he had to sell his father's farms. This seems to have reformed him, for he then moved to London to work with A. W. Hofmann at the Royal College of Chemistry, where he was readily noticeable by his German semi-military dress. He is reported to have worn the largest top hat ever seen in Oxford Street. He then went to work as a chemist in Allsopp and Son's brewery in Burton-on-Trent, carrying out research into azo dyes in his spare time. He longed to return to Germany and carry out further research in chemistry, but this was not to be since he died in 1888 while on holiday in Bournemouth.

Figure 5.7
Johann Peter Griess 1829–1888.

5.2 Coupling reactions of diazonium salts

Diazonium cations are electrophilic reagents in their own right, and can take part in aromatic substitution reactions with other aromatic compounds. When they do so, the result is a **coupling** of two aromatic rings via the two nitrogen atoms. Two of the simplest coupling reactions to give azo dyes are:

diazonium salt   coupling component   azo dye

(5.7)

para-hydroxyphenylazobenzene

phenolate anion

N,N-dimethylaniline

(5.8)

para-dimethylaminophenylazobenzene

◔ Write out the mechanism for the reaction of the electrophilic diazonium cation with the phenolate ion to generate *para*-hydroxyphenylazobenzene, as shown above (don't worry about drawing out all the resonance forms!).

◑ The diazonium cation, $Ar-\overset{+}{N}\equiv N$ *, which acts as an electrophile, is attacked by the aromatic ring of the phenolate ion to form an intermediate cation, which then loses a proton to give an anion; this is then re-protonated on work-up to give the product:

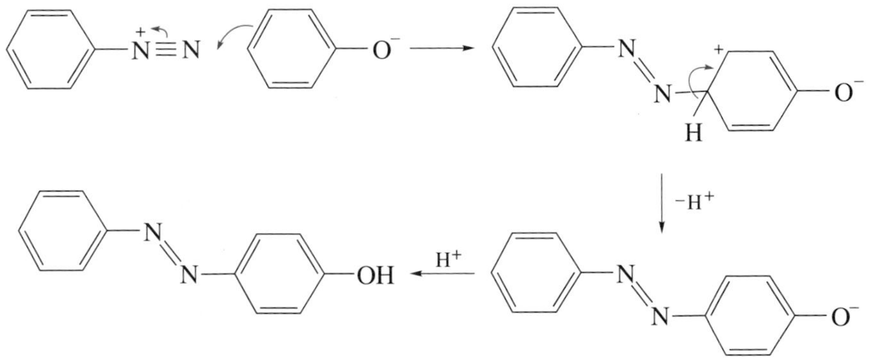

SCHEME 5.2

5.3 Substitution reactions of diazonium salts

Substitution reactions of diazonium salts are the best general way of introducing F, Cl, Br, I, CN, OH and H into an aromatic ring; the range of reactions is shown in Scheme 5.3. Diazonium salts are valuable in synthesis, not only because they react to form so many classes of compound, but because they can be prepared from nearly all aromatic amines. There are few types of substituent whose presence in a molecule interferes with diazotization. The amines from which diazonium compounds are prepared are readily obtained from the corresponding nitro compounds, which are themselves prepared by direct nitration. Diazonium salts are thus an important link in the synthesis of a variety of aromatic compounds:

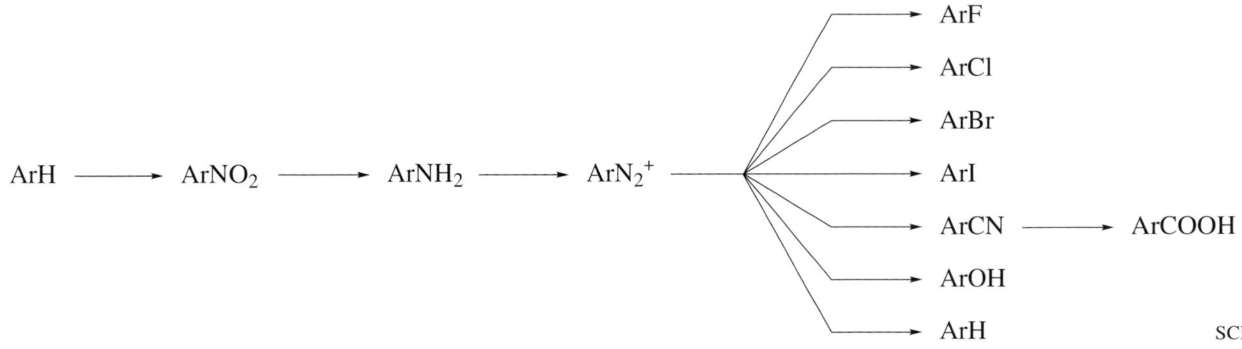

$$ArH \longrightarrow ArNO_2 \longrightarrow ArNH_2 \longrightarrow ArN_2^+ \longrightarrow$$

ArF
ArCl
ArBr
ArI
ArCN $\longrightarrow$ ArCOOH
ArOH
ArH

SCHEME 5.3

We shall not discuss the mechanisms of these reactions. Instead, the reagents that are used for each of them are summarized in the Appendix, which draws together all the synthetically useful reactions contained in Part 2. These reactions also appear in the *Data Book* (available from the CD-ROM).

---

* Just as R signifies an alkyl group, the aromatic equivalent is often denoted as 'Ar'.

Notice that reaction via a diazonium salt provides a way of introducing both iodine and fluorine atoms into an aromatic ring, which is something that cannot be done by direct halogenation (Section 3.3). Also, note the replacement of the diazonium group by hydrogen, which provides a way of removing an $-NH_2$ or $-NO_2$ group from an aromatic ring.

5.4 Summary of Section 5

1 Aromatic diazonium salts are prepared by treating cold acidic solutions of aromatic amines with sodium nitrite. The solutions are generally not isolated, but are prepared and used immediately in the required synthesis.

2 The coupling of diazonium salts with aromatic phenols or amines yields azo compounds.

3 Many useful reactions are known in which a diazonium cation is converted into product with the loss of nitrogen gas. These substitution reactions, together with the required reagents, are summarized in the Appendix.

QUESTION 5.1

What are the formulae of the compounds A–H in reaction sequences (a) and (b)?

(a) benzene $\xrightarrow{HNO_3/H_2SO_4}$ A $\xrightarrow{Fe/HCl}$ B $\xrightarrow[0\,°C]{NaNO_2/HCl}$ C $\xrightarrow{C_6H_5OH}$ D

(b) nitrobenzene $\xrightarrow{Br_2/FeBr_3}$ E $\xrightarrow{Fe/HCl}$ F $\xrightarrow[0\,°C]{NaNO_2/HCl}$ G $\xrightarrow{CuBr}$ H

QUESTION 5.2

(a) Write down the mechanism of the following diazotization reaction:

$$^-O_3S-\!\!\!\bigcirc\!\!\!-NH_2 \xrightarrow[0\,°C]{NaNO_2/HCl} {}^-O_3S-\!\!\!\bigcirc\!\!\!-\overset{+}{N}\equiv N \qquad (5.9)$$

(b) Write down the mechanism of the following coupling reaction to yield substitution at the 2 and 4 positions in the phenol ring:

$$\underset{\text{phenol}}{\overset{OH}{\bigcirc}} + N\equiv\overset{+}{N}-\!\!\!\bigcirc\!\!\!-SO_3^- \longrightarrow ? \qquad (5.10)$$

APPENDIX   SUMMARY OF REACTIONS USEFUL IN THE SYNTHESIS OF AROMATIC COMPOUNDS

The direct monosubstitution of benzene

The formation of a diazonium salt

Substitution reactions of a diazonium salt

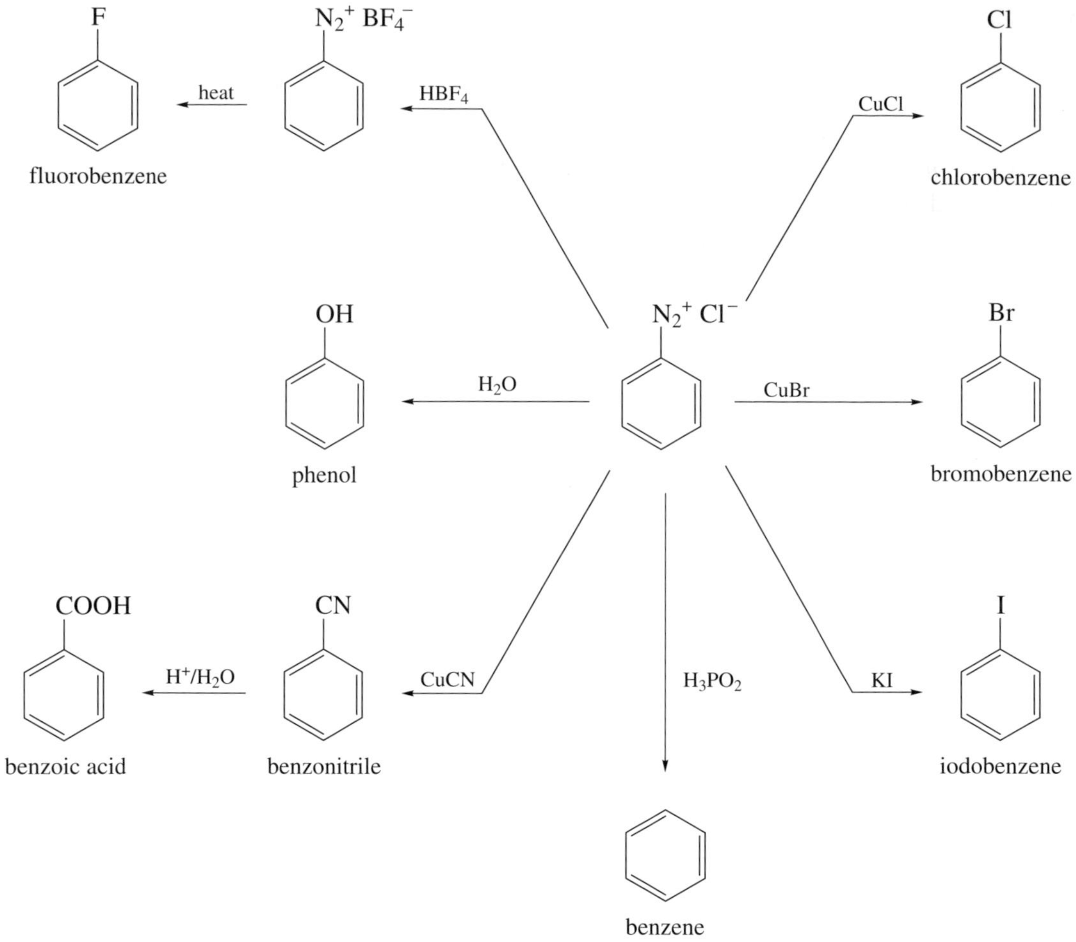

fluorobenzene

chlorobenzene

phenol

bromobenzene

benzoic acid

benzonitrile

iodobenzene

benzene

Coupling reaction of a diazonium salt

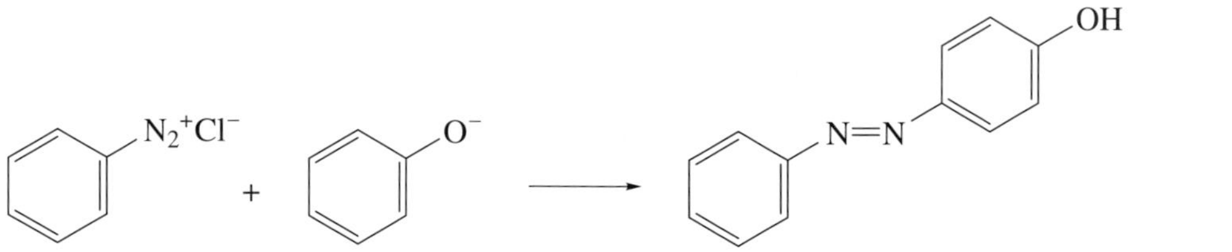

para-hydroxyphenylazobenzene (an azo dye)

LEARNING OUTCOMES

Now that you have completed *Alkenes and Aromatics: Part 2 — Aromatic compounds*, you should be able to do the following things:

1   Recognize valid definitions of, and use in a correct context, the terms, concepts and principles in the following Table. (All Questions)

List of scientific terms, concepts and principles introduced in *Aromatic compounds*

Term	Page number	Term	Page number
activation	57	Friedel–Crafts acylation	58
acylonium ion	58	Friedel–Crafts alkylation	54
aromatic compound	40	halogenation	50
aromaticity	43	Lewis acid	50
Clemmensen reduction	59	*meta* isomer	78
coupling reaction	72	nitration	48
deactivation	58	nitronium ion	49
delocalized π-electron system	44	*ortho* isomer	78
		para isomer	78
diazotization	72	polyalkylation	58
directing effect	62	rearrangement	56
electrophilic aromatic substitution reaction	45	sulfonation	52

2   Explain how the concept of resonance helps in the understanding of both the structure and the stability of the benzene molecule. (Questions 2.1 and 3.1)

3   Draw the positional isomers of various substituted derivatives of benzene. (Question 2.2)

4   Explain the mechanism of electrophilic aromatic substitution, with reference to the nitration, halogenation, sulfonation, alkylation and acylation of benzene. (Questions 3.4, 3.5, 3.6 and 3.7)

5   Depict the course of an electrophilic aromatic substitution reaction by means of a reaction-coordinate diagram. (Questions 3.3 and 3.4)

6   Draw out the resonance forms for the intermediate formed in an electrophilic aromatic substitution reaction, and use them to explain the activating and directing effects of a substituent on a benzene ring. (Question 3.2)

7   Use a knowledge of directing effects to predict the position of substitution in certain reactions of monosubstituted benzene derivatives. (Questions 4.1, 4.2 and 4.3)

8   Predict the products or reagents for both the coupling and substitution reactions of diazonium salts. (Questions 5.1 and 5.2)

QUESTIONS: ANSWERS AND COMMENTS

QUESTION 2.1 (*Learning Outcome 2*)

The representation:

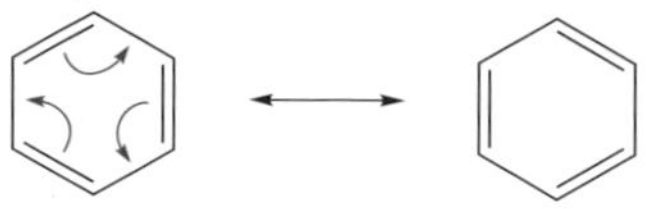

shows two skeletal structures joined by a double-headed arrow. This arrow is not an equilibrium arrow; it is a resonance arrow, indicating that the overall structure is part way between the two extremes shown above. *It does not mean that benzene spends some time in one form and some time in the other*. The bonding between any two adjacent carbons is a single bond in one resonance form and a double bond in the other. This means that in the actual structure there is effectively one and a half bonds between each adjacent pair of carbons, and that all bond lengths are the same. The curly arrows show the electron movement that converts one resonance form into the other. Again, this does not mean that the electrons are resonating between these structures. In the actual structure there is an overall symmetrical distribution of electrons between the six carbon–carbon bonds.

QUESTION 2.2 (*Learning Outcome 3*)

(a) Assuming a perfectly regular hexagonal structure for benzene, there are three possible structural isomers of $C_6H_4Cl_2$:

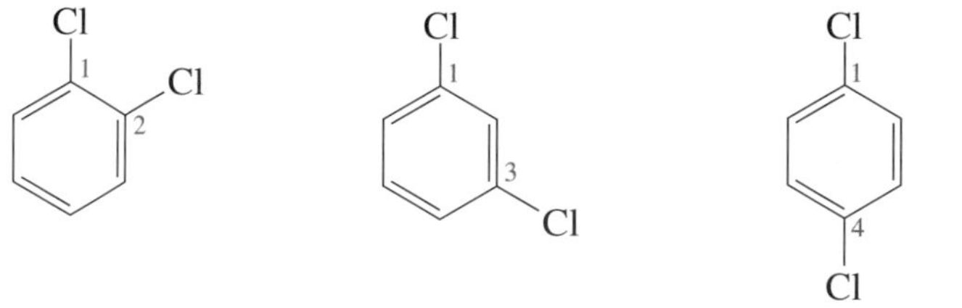

1,2-dichlorobenzene	1,3-dichlorobenzene	1,4-dichlorobenzene
ortho **Q.1**	*meta* **Q.2**	*para* **Q.3**

There is only one 1,2-dichlorobenzene (*ortho* isomer), since

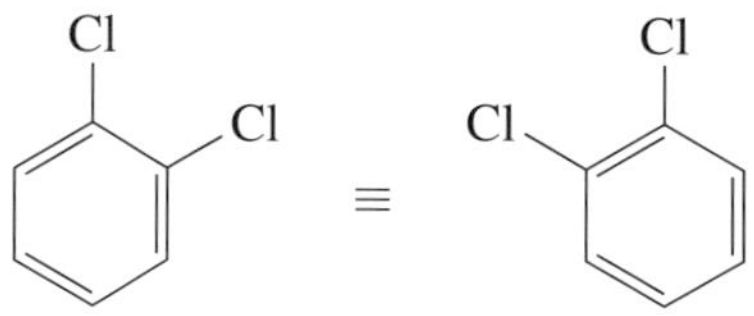

Furthermore, there is only one possible 1,3-isomer (*meta* isomer) and one possible 1,4-isomer (*para* isomer). Isomers that arise from substitution at different positions on the benzene ring are known as *positional isomers*.

(b)   If we assume that localized bonds are present in benzene, there would be alternate long single and short double bonds around the ring; in other words, with such a structure the ring would no longer be a regular hexagon. The number of possible disubstituted isomers would now rise to four:

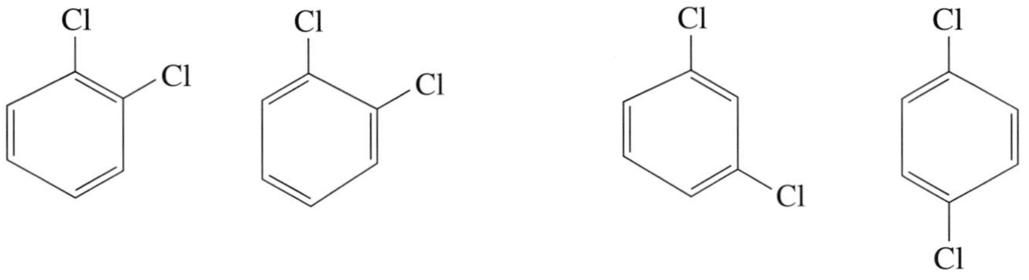

| ortho | ortho | meta | para |

The 'extra' 1,2 (*ortho*)-isomer arises because a double bond is shorter than a single bond. However, we know that in reality only one 1,2-dichlorobenzene is known. This is only explicable in terms of a fully delocalized structure for benzene.

QUESTION 3.1 (*Learning Outcome 2*)

If positive charges are adjacent to double bonds, they can be stabilized by resonance*. The carbocation generated by attack of the electrophile on the benzene ring has a positive charge adjacent to a carbon–carbon double bond, so it can be stabilized by resonance as shown below.

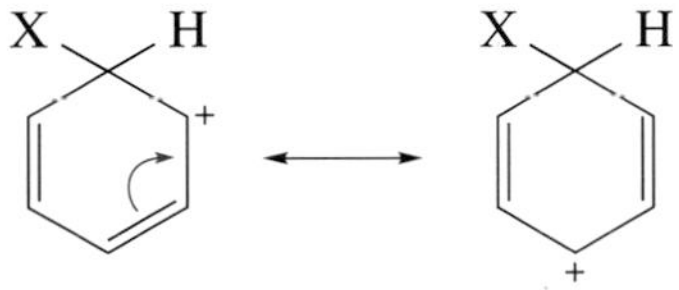

In fact another resonance form can be generated by moving the electrons in the left-hand double bond:

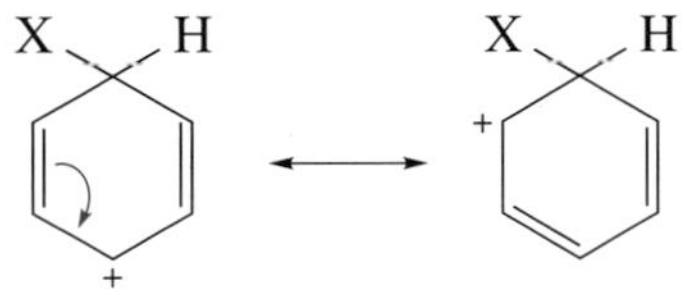

This gives a total of three resonance forms:

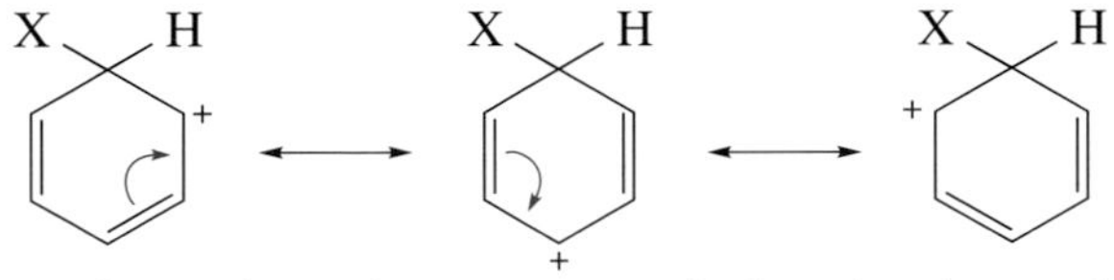

If we refer to the carbon atom attached to the electrophile (X) as carbon 1, then the positive charge is shared between carbon atoms 2, 4 and 6; each of these atoms carries about a third of a positive charge. Regardless of the nature of X, the charge on the ring is spread over these three carbon atoms, which results in extra stabilization.

---

* Resonance is the subject of a program on the CD-ROM with *Chemical Kinetics and Mechanism*[3].

QUESTION 3.2 (Learning Outcome 6)

The resonance forms for the carbocations formed in this electrophilic aromatic substitution are shown below:

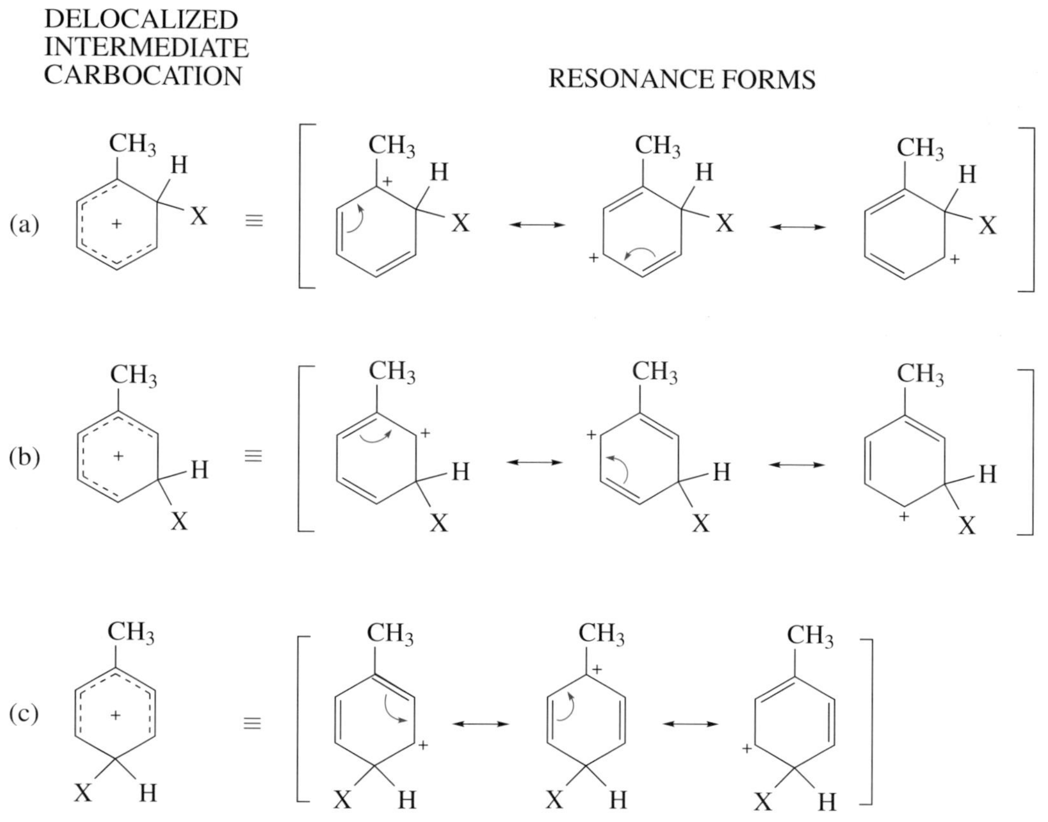

DELOCALIZED
INTERMEDIATE
CARBOCATION   RESONANCE FORMS

(a)

(b)

(c)

QUESTION 3.3 (Learning Outcome 5)

The complete diagram is shown in Figure Q.1.

(a) The positions of the activated complexes are marked with a double dagger, ‡

(b) E_1 is the activation energy (E_a), which is the energy barrier that must be overcome in forming the intermediate carbocation. This is the slow, rate-limiting step, and hence it is this energy difference that determines the rate of the reaction.

(c) E_2 is the difference in energy between the reactants and products. It is this energy difference that is related to the enthalpy change, and thus the Gibbs function change, $\Delta G_m^{\ominus}$, and the size of the equilibrium constant, and therefore to the extent of the reaction at equilibrium.

QUESTION 3.4 (Learning Outcomes 4 and 5)
The complete diagram is shown in Figure Q.2.

QUESTION 3.5 (Learning Outcome 4)

The mechanism can be shown as follows:

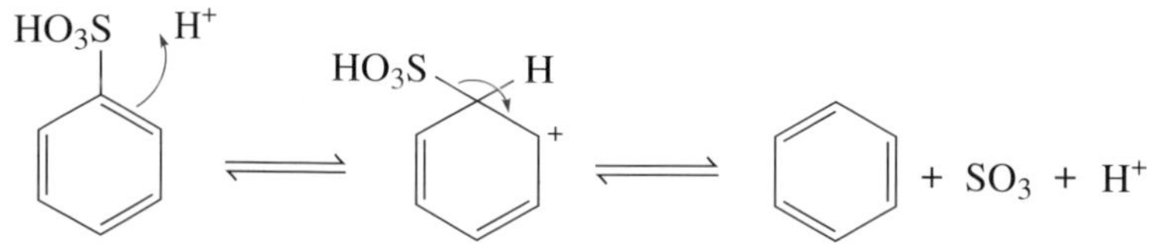

$$+ \; SO_3 + H^+ \qquad\qquad (Q.1)$$

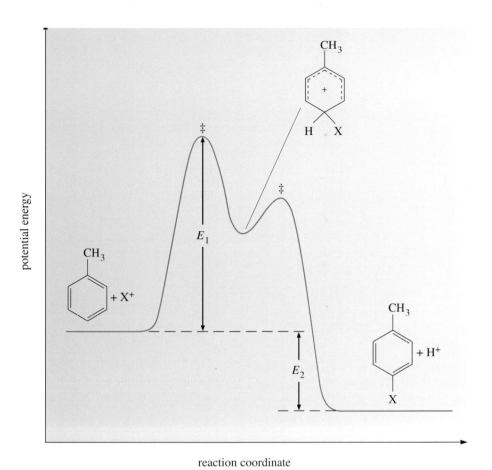

Figure Q.1
Reaction-coordinate diagram for the *para* substitution of methylbenzene.

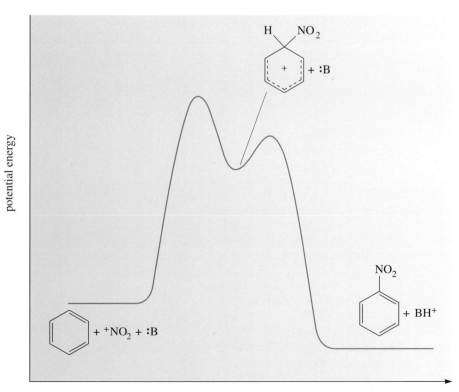

Figure Q.2
Reaction-coordinate diagram for the nitration of benzene.

Don't worry too much about the actual species lost in the second step. It could be a concerted breaking down of the SO_3H group:

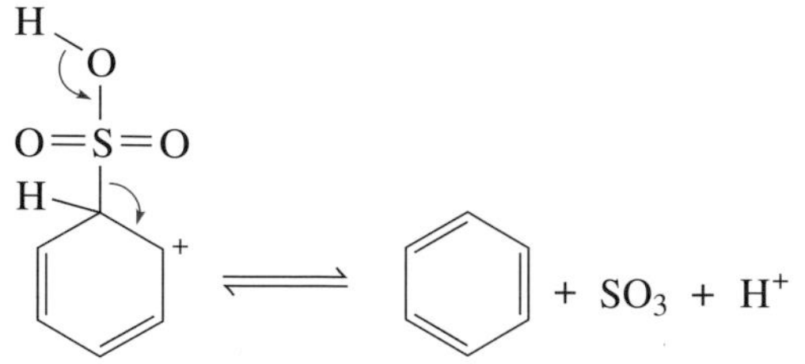

$$+ SO_3 + H^+ \qquad\qquad (Q.2)$$

or perhaps more likely, the SO_3 could be formed after loss of a proton:

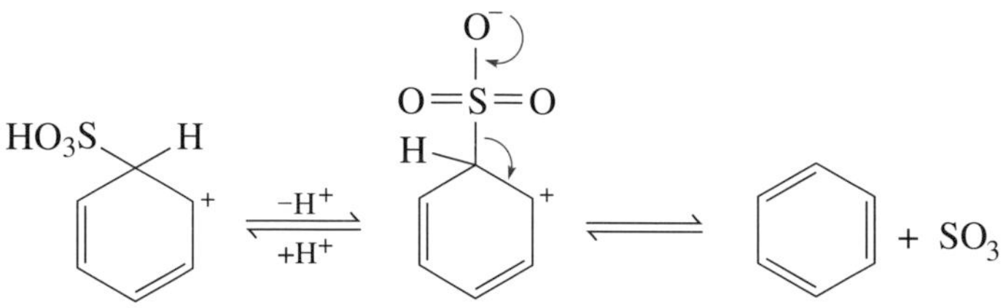

$$+ SO_3 \qquad\qquad (Q.3)$$

QUESTION 3.6 (*Learning Outcome 4*)

Locating the charge on the $\overset{+}{N}O_2$ on the central nitrogen gives a resonance form in which the nitrogen has four valence electrons. Each oxygen provides two electrons to form a double bond, making eight valence electrons in all. These will be distributed as four electron pairs in the two double bonds to oxygen. There are therefore only two repulsion axes, and therefore $\overset{+}{N}O_2$ is linear $O=\overset{+}{N}=O$.

QUESTION 3.7 (*Learning Outcome 4*)

Both the products in parts (a) and (b) are the result of Friedel–Crafts alkylations, which can be described by the following general reaction scheme:

$$RCH_2Cl + AlCl_3 \rightleftharpoons R\overset{+}{C}H_2 + AlCl_4^-$$

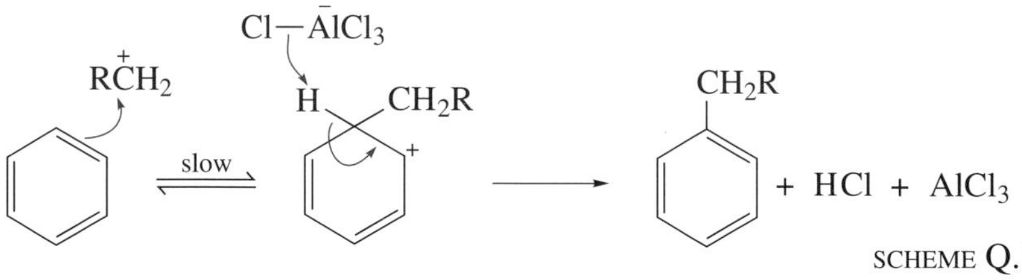

$$+ HCl + AlCl_3$$

SCHEME Q.1

The carbocation can undergo rearrangement; this appears to be the situation in both (a) and (b):

(a) $\quad CH_3CH_2CH_2Cl \xrightarrow{AlCl_3} CH_3CH_2\overset{+}{C}H_2 \rightleftharpoons CH_3\overset{+}{C}HCH_3$

(b) $\quad CH_3CH_2CH_2CH_2Cl \xrightarrow{AlCl_3} CH_3CH_2CH_2\overset{+}{C}H_2 \rightleftharpoons CH_3CH_2\overset{+}{C}HCH_3$

In each case, rearrangement occurs to form the more-stable secondary carbocation. The product distribution indicates that although some of the primary carbocation is present, the major product results from electrophilic attack by the secondary carbocation.

In reaction (c), benzene and propanoic anhydride produce the ketone 1-phenyl-propan-l-one. The electrophile that reacts with benzene is the acylonium ion, $CH_3CH_2-\overset{+}{C}=O$, generated by reaction between the anhydride and $AlCl_3$. The detailed mechanism is shown in Scheme Q.2.

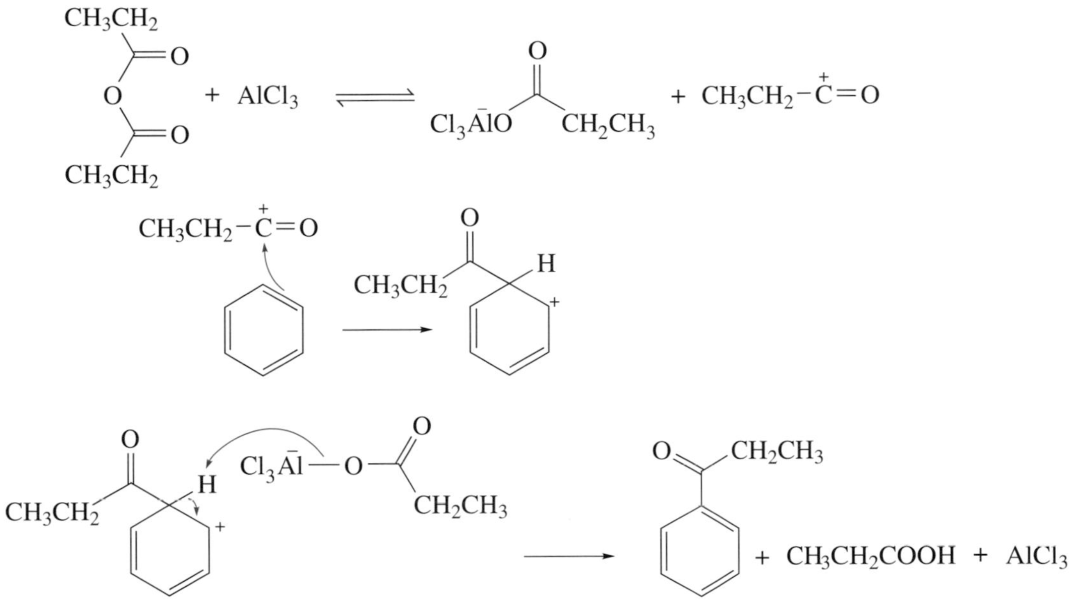

SCHEME Q.2

QUESTION 4.1 (Learning Outcome 7)

(a) The three reactions with methoxybenzene are:

(i)

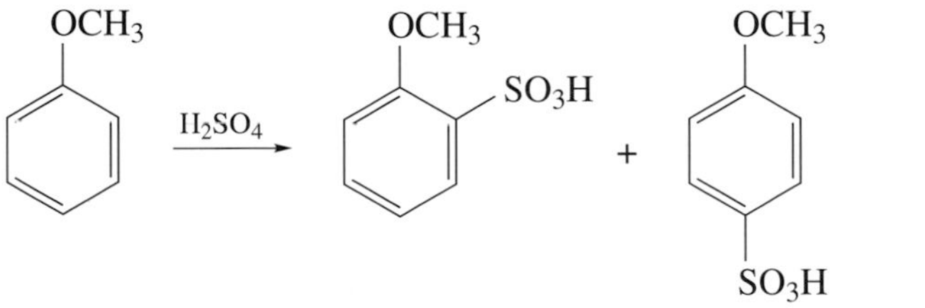

(ii)

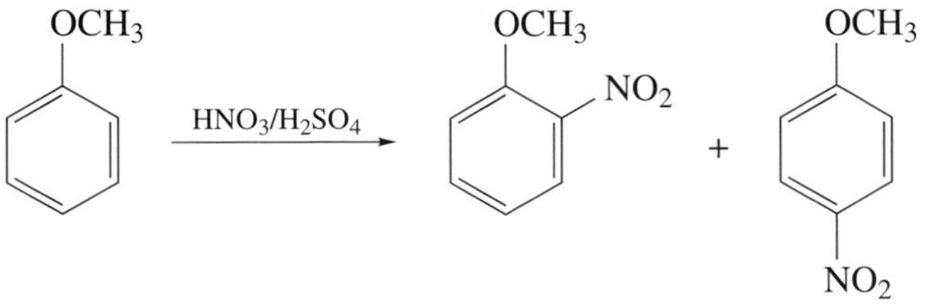

(iii)

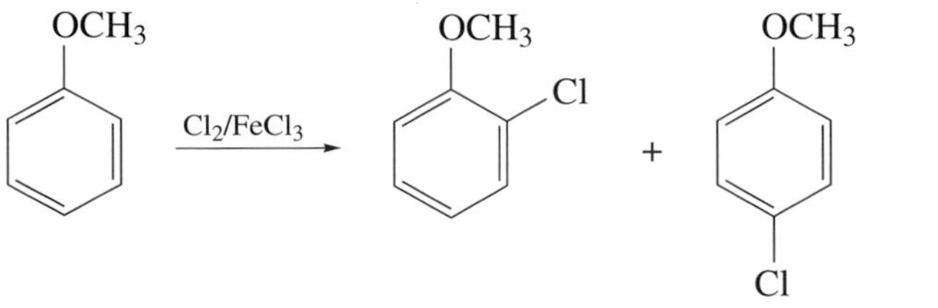

(b)   The three reactions with nitrobenzene are:

(i) H_2SO_4

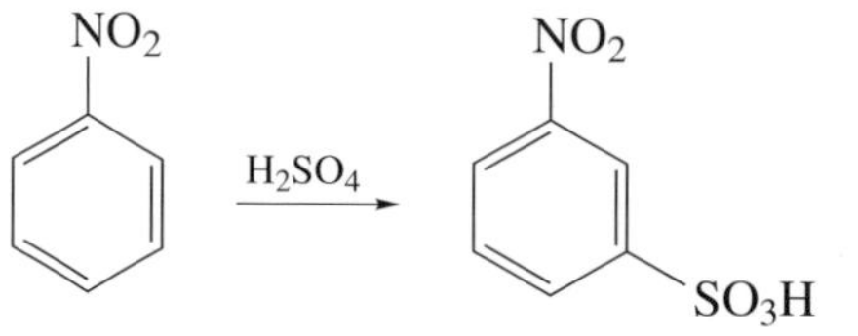

(ii) HNO_3/H_2SO_4

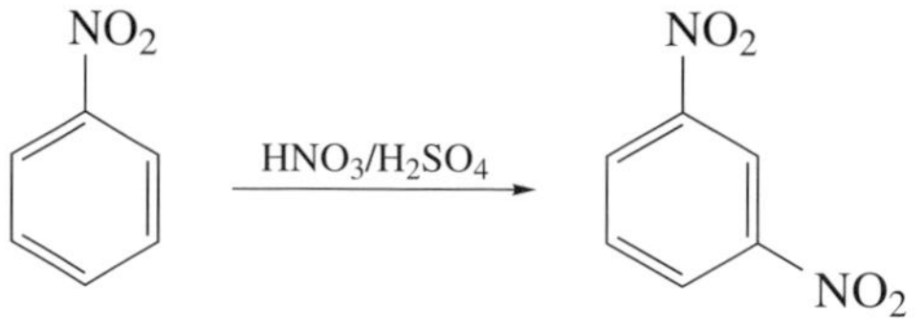

(iii) $Cl_2/FeCl_3$

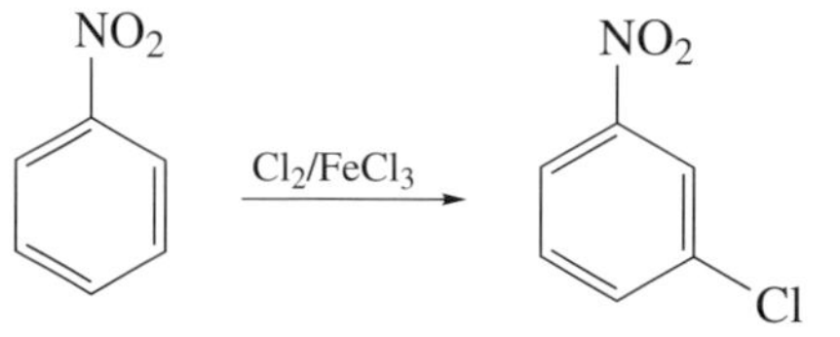

QUESTION 4.2 (*Learning Outcome 7*)

The reagents are listed in Table Q.1. Note that in each example the required product will need to be isolated where *ortho* and *para* isomers are both major products.

Table Q.1

Reaction	First step	Second step	Comment
(a)	$Cl_2/FeCl_3$	HNO_3/H_2SO_4	Cl group *ortho/para*-directing
(b)	HNO_3/H_2SO_4	$Cl_2/FeCl_3$	NO_2 group *meta*-directing
(c)	$CH_3COCl/AlCl_3$	$Cl_2/FeCl_3$	CH_3CO group *meta*-directing
(d)	$CH_3CH_2Cl/AlCl_3$	H_2SO_4	CH_3CH_2 group *ortho/para*-directing

QUESTION 4.3 (*Learning Outcome 7*)

In each of the examples (a)–(c) one of the aromatic compounds is activated to electrophilic substitution under the reaction conditions, and the other is deactivated. In each case, the more-activated compound contains an *ortho/para*-directing substituent. We therefore expect that the activated compound will react faster, and that the major products will be *ortho*- and *para*-substituted derivatives of the activated compound.

(a) The major products will be

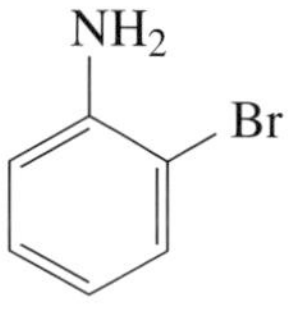

and

The NH_2 group is *ortho/para*-directing and activating, whereas the CH_3CO group is deactivating and *meta*-directing.

(b) As the CH_3 group is *ortho/para*-directing and activating, and the NO_2 group is *meta*-directing and deactivating, the major products will be

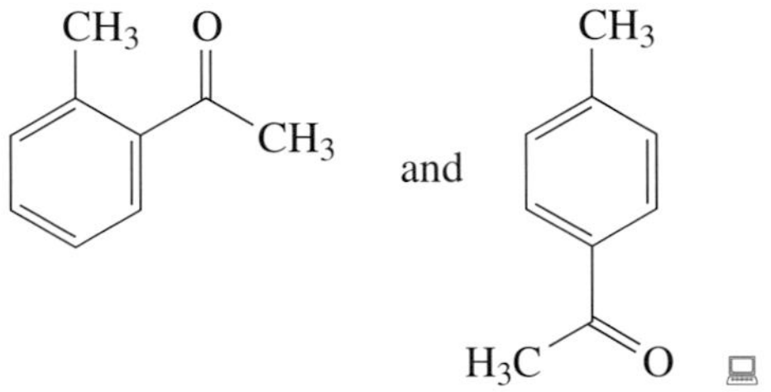

and

(c) You may have found this part difficult because both OCH_3 and NH_2 are *ortho/para*-directing and activating. However, consider the reaction conditions and the basic nature of the amino group. Under the conditions of the reaction the following transformation takes place:

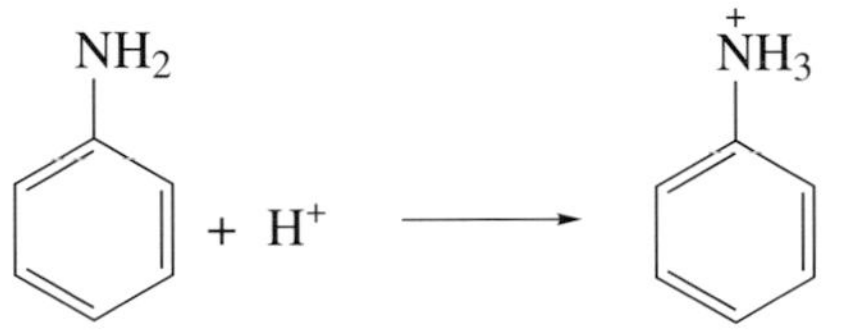

So the real comparison is between the *ortho/para*-directing and activating OCH_3 group, and the $\overset{+}{N}H_3$ group, which is deactivating and *meta*-directing. The major products are therefore

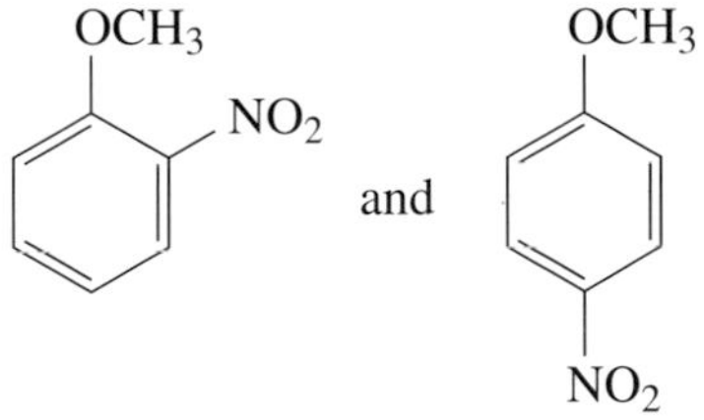

and

QUESTION 5.1 (*Learning Outcome 8*)

(a)

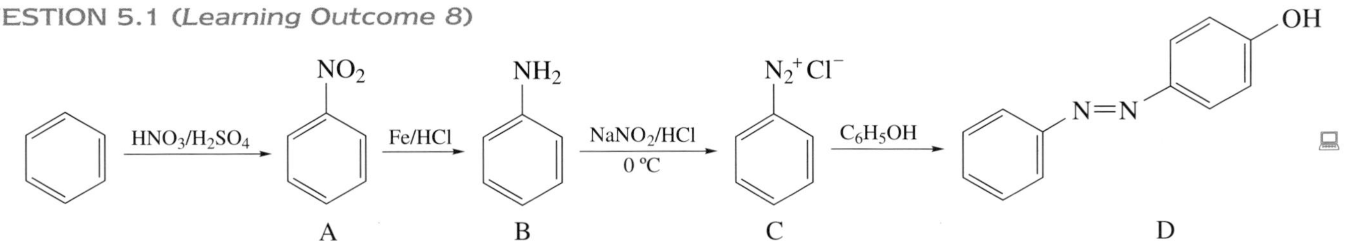

(b)

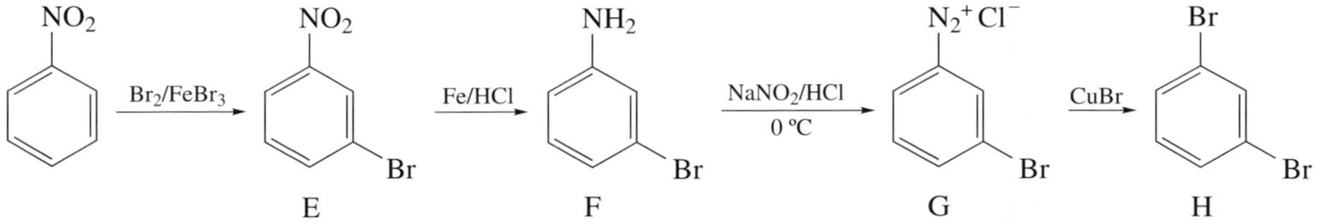

QUESTION 5.2 (*Learning Outcome 8*)

(a)   The mechanism for diazotization is:

$$H-\overset{..}{\underset{..}{O}}-N=O \quad \rightleftharpoons \quad \overset{H}{\underset{H}{\overset{|}{\underset{|}{O}}}}{}^{+}-N=O \quad \rightleftharpoons \quad \overset{+}{N}=O + H_2O \qquad \text{generation of the electrophile}$$

attack of the electrophile   loss of a proton   rearrangement

loss of water   protonation

(b) The mechanism for coupling at the 2 and 4 positions in phenol is:

COUPLING AT THE 2-POSITION

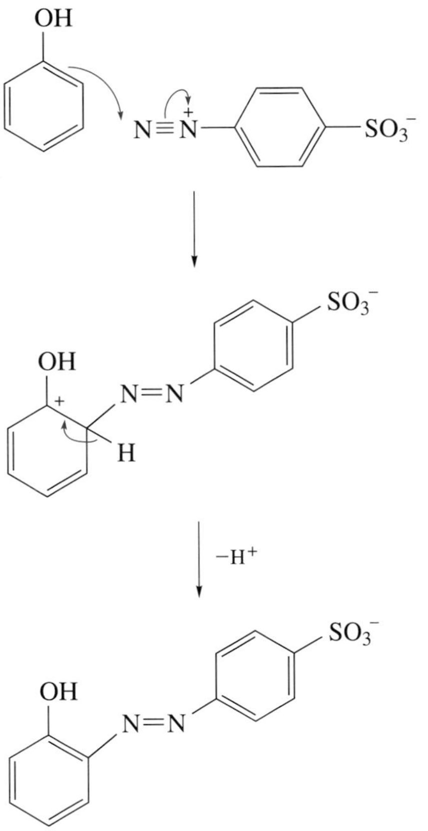

COUPLING AT THE 4-POSITION

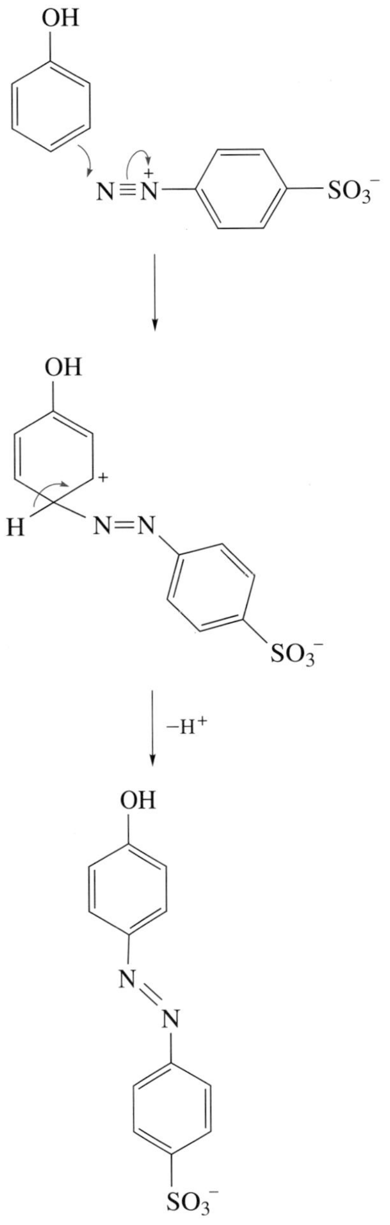

The carbocation intermediates in both 2-substitution and 4-substitution are stabilized by extra resonance forms involving the hydroxyl oxygen. You should convince yourself of this by drawing out the resonance forms in each case (cf. Structure **4.7**, p. 63).

FURTHER READING

1   E. A. Moore (ed.), *Molecular Modelling and Bonding*, The Open University and the Royal Society of Chemistry (2002).

2   P. G. Taylor (ed.), *Mechanism and Synthesis*, The Open University and the Royal Society of Chemistry (2002).

3   M. Mortimer and P. G. Taylor (eds), *Chemical Kinetics and Mechanism*, The Open University and the Royal Society of Chemistry (2002).

ACKNOWLEDGEMENTS

Grateful acknowledgement is made to the following source for permission to reproduce material in this part:

Figure 1.1: Institute of Gas Engineers; *Figure 1.2*: Museum of Science and Industry in Manchester; *Figure 1.4*: Adelheid Raqué-Nuttall; *Figures 2.1, 3.2 and 5.7*: reproduced by courtesy of the Library and Information Centre, Royal Society of Chemistry; *Figure 3.4*: courtesy of MIT Museum, Massachusetts, USA; *Figures 5.2 and 5.6*: Science Museum/Science and Society Picture Library; *Figure 5.5*: courtesy of University Pennsylvania Library.

Every effort has been made to trace all the copyright owners, but if any has been inadvertently overlooked, the publishers will be pleased to make the necessary arrangements at the first opportunity.

Part 3

A First Look at Synthesis

edited by Michael Gagan

based on *The search for new drugs*
by Alan Bassindale (1991)

and *Making drugs*
by Peter Taylor (1991)

STRATEGY FOR THE DISCOVERY OF NEW DRUGS

So far in this Book, we have concentrated on describing single reactions, which convert one class of organic compound into another. We have looked at the reagents needed to bring about reaction, the substrates that undergo the reaction, and the conditions under which reactions take place. We have also discussed ideas of reaction mechanism — how reactions take place at the molecular level. However, there has been a 'hidden agenda' behind our study of individual reactions. We have been building up a toolkit of effective procedures that could also be utilized as steps in a grander scheme for the construction of a desirable, but complex, organic molecule — in short, preparing for **synthesis**.

All types of reaction may be utilized in synthesis. *Substitution* of a halogen group by a nucleophile, like ammonia or an amine, can be used to introduce a nitrogen function; and replacing a halogen with cyanide anion, ^-CN, will add one more carbon atom to a molecule. *Elimination* of H—X, such as water (H—OH) from an alcohol, or H—Br from a bromoalkane, will introduce a carbon–carbon double bond into the skeleton; and *addition* of X_2 or HX to a double bond allows us to introduce a substitutable halogen group into the carbon chain.

🔵 How could you attach a carbon chain, containing a reactive functional group, to a benzene ring?

⚪ An *electrophilic aromatic substitution*, like the Friedel–Crafts acylation reaction, would attach a carbon chain with a carbonyl group adjacent to the ring.

Perhaps the most conspicuous successes of the synthetic chemist have been in the field of pharmaceuticals, in the preparation of drugs. A common scientific definition of a drug is 'a chemical useful in the therapeutic treatment of disease, or in clinical practice'. Herbal (or folk) medicine has been practised since the earliest civilizations, perhaps even from prehistoric times. In Britain in the fourteenth century, we know from Chaucer that drugs were in common use. He writes in the Prologue to *The Canterbury Tales*, of the 'Doctour of Physik':

> He was a perfect practising physician.
> He gave the man his medicine then and there.
> All his apothecaries in a tribe
> Were ready with the drugs he would prescribe ...

This is Neville Coghill's translation (Penguin Books, 1951). The original transliteration is:

> He was a verray, parfit practisour.
> The cause yknowe, and of his harm the roote,
> Anon he yaf the sike man his boote.
> Ful redy hadde he his apothecaries
> To send him drogges and his letuaries*

---

* boote = prescription; letuary = cure.

Shakespeare also mentions drugs, most notoriously the death-simulating drug that resulted in the untimely death of Romeo, and subsequently Juliet.

Although many of the cures and potions described in the early herbals were unreliable, some were remarkably effective. Chewing willow bark is no longer our choice, but this folk remedy led to the development of the world's most popular drug, aspirin (**1.1**). The compound digitalin (**1.2**; R is a complex side-chain), used to control irregular heartbeat, is still isolated from an effusion of dried foxgloves (*digitalis*), a potion used for centuries in the treatment of dropsy (heart failure).

Today, a very large majority of the drugs on which we rely for prevention and cure of our bodily ailments, are synthetic materials. Drugs have been spectacularly successful in reducing mortality and morbidity resulting from infectious diseases caused by bacteria. In particular, the penicillins (**1.3**), sulfonamides (**1.4**) and other antibiotics have helped to make good health commonplace in the industrialized world, and are generally considered to be an essential part of the fight against disease in developing countries.

The positive side of drug therapy is occasionally overlooked because of concern about the negative effects of drug misuse. Such concern should not overshadow achievements such as the dramatically reduced number of deaths from tuberculosis and pneumonia, both of which were major causes of death until the middle of the twentieth century. Many bacterial infections have successful chemical antidotes, but the search for drugs to combat cancers and viral diseases has turned out to be much more difficult. Some major successes have recently appeared, with antiviral research being stimulated by the fight to control HIV and AIDS. A deeper understanding of biochemical pathways, and of the structure and activity of enzymes and receptors (Section 1.1), has also allowed the development of drugs to treat other medical conditions.

1.1

1.2

1.3

1.4

Captopril for high blood pressure

zidovudine (AZT) for HIV

Ranitidine (Zantac) for gastric ulcers

cisplatin for cancer

Unfortunately, bacteria have the ability to develop resistance to heavily administered drugs, and some antidotes that were previously used successfully have now become ineffective against resistant strains. HIV also has the ability to adapt its enzymes to be unresponsive to the drug, while continuing to carry out their function with their normal substrates. So the continuing quest for more potent drugs with fewer side-effects presents an exciting challenge for the drug synthesisers of the future.

The traditional ways of searching for new drugs were often based on folk medicine. The first task was to isolate the active ingredient from the mixture of compounds present in the plant extract. The structure and stereochemistry of this molecule was then determined. Various model compounds would then be synthesised, in which each one contained a deliberate change to the natural molecule. They would all be tested for efficacy in an attempt to identify the parts of the structure that were responsible for the pharmaceutical activity — that is, to determine **structure–activity relationships**. The results were then used to try to predict the optimum molecular structure for achieving the highest activity. This process, which usually required the use of animals to test potentially effective compounds, was very laborious.

Today a more rational approach is taken to drug design. Using a detailed knowledge of the chemistry and biochemistry of normal and diseased organisms, molecules can be constructed with the specific purpose of aiding or blocking a particular biochemical pathway. Complete structures of some proteins including their 'receptors' (sites on a large molecule where a smaller molecule can bind to produce a physiological effect) have now been determined by techniques such as X-ray crystallography, sometimes with the natural substrate bound to it[*]. When this information is available, the structure can be displayed on a computer screen, and the shape, and other physical characteristics of the active site can be determined. The substrate may then be deleted electronically, and replaced with a potential drug molecule. If a good 'fit' is obtained, that compound can be synthesised and tested[†].

Two other developments have enormously speeded up the process of drug synthesis and testing. The first, **combinatorial chemistry**, is a way of simultaneously making dozens, even hundreds, of molecules with slightly different structures in an automated procedure. However, this approach would not have been so successful without the parallel development of microscale, rapid-throughput testing procedures, based mainly on bioassay techniques, using protein extracts rather than whole animal tests. Once a promising candidate has been identified, it can be prepared in larger quantities for full-scale testing.

So after this brief introduction to drug design and synthesis, we shall now take a break from the study of individual reactions. We shall examine a group of drugs that operates on the central nervous system, and describe some of the factors that must be considered when making (or *synthesising*) such compounds.

1.1 Neurotransmitters and receptors

The part of the human nervous system which controls involuntary action such as heartbeat, is called the *autonomic system*. The section of the autonomic nervous system that keeps the body 'ticking over' in its normal working rhythm is called the *parasympathetic system*.

The other part of the autonomic nervous system that we are interested in is the *sympathetic system*. This comes into play particularly when we are under stress. The body is said to be prepared for 'fight, flight or frolic'. Additionally, the sympathetic system, in conjunction with the parasympathetic system, is sometimes used to

---

[*] The use of X-ray crystallography for the determination of crystal structure is discussed in Part 1 of *The Third Dimension*[1].

[†] More detail on the techniques for modelling drug structure is given in the Case Study in *Molecular Modelling and Bonding*[2].

maintain the balance of particular pairs of muscles. In the eye, for example, the pupil size is balanced and determined by the parasympathetic and sympathetic systems working in opposition.

Before any muscle (including the heart) can act, a message must be sent to that muscle telling it to contract. These messages are carried by compounds known as **neurotransmitters**, or 'chemical mediators'. The neurotransmitter for the parasympathetic system is acetylcholine (2-trimethylammonioethyl ethanoate, **1.5**).

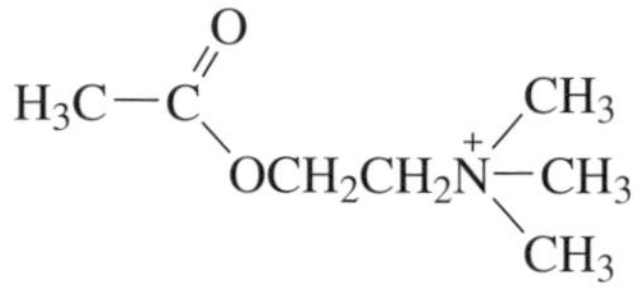

⬤ What are the functional groups in acetylcholine?

⬤ There are two functional groups, an ester (**1.6**) and a quaternary ammonium group (**1.7**).

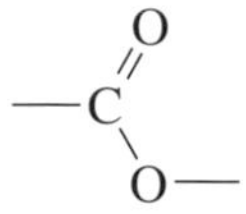

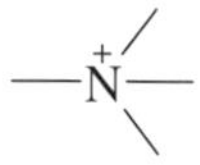

The acetylcholine molecule bears an overall positive charge. Electrical neutrality is maintained by an equal number of anions, such as Cl^- or CH_3COO^-.

The generally accepted explanation for how the acetylcholine controls the excitation of muscles, is that the molecule attaches itself to specific receptors within the nervous system. Neurotransmitter or drug molecules are usually bound to the receptor by at least two different interactions between separate parts of the molecule and neighbouring sites in the receptor (Figure 1.1). Binding at more than one site ensures that only molecules with appropriate geometries and functional groups can fit the receptor. Complex changes occur when the acetylcholine becomes bound, and these stimulate muscle contraction.

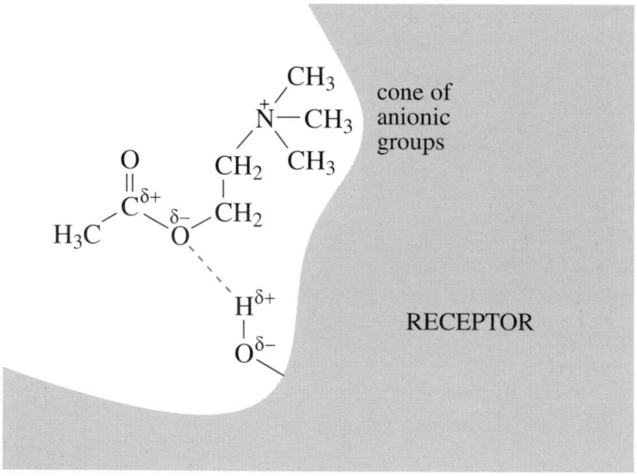

Figure 1.1 Acetylcholine bound to a receptor site.

However, in the heart section of the sympathetic nervous system there are two different neurotransmitters working at different places in the chain of command. One is acetylcholine and the other is noradrenaline (**1.8**). The prefix 'nor-' is not part of the IUPAC system of nomenclature. In trivial names it denotes a compound with one CH_2 less than the parent compound. Thus, noradrenaline has a hydrogen attached to the nitrogen, whereas adrenaline has a methyl group in that position.

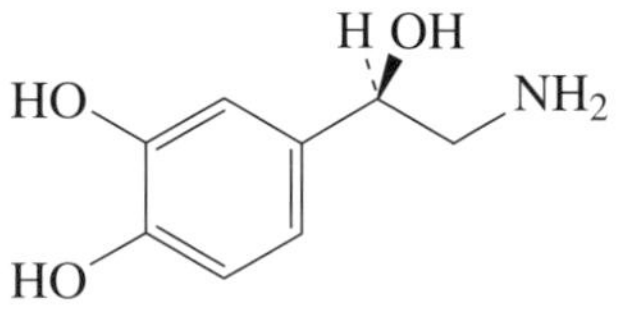

⬤ What are the main functional groups in noradrenaline?

⬤ There is a benzene diol (called a *catechol*), a secondary alcohol, and a primary amine, RNH_2.

Let us first look at how a neurotransmitter molecule binds to the receptor. It is very unusual to find that neurotransmitters (and drugs) are bound to receptor sites by two-electron covalent bonds. Covalent bonds, such as C—C, C—O or C—N single bonds are very strong, and would require a large amount of energy to break them once formed. The neurotransmitters are held on to the receptor by one or more of the familiar types of *intermolecular force**. Because these interactions are weak, reversible binding can be achieved with a minimum expenditure of energy.

● Suggest some of the intermolecular forces that might help bind a molecule to a receptor surface.

● Hydrogen bonds, electrostatic interactions, London forces. A combination of any or all of these interactions contributes to the binding.

Although each of these interactions is relatively weak (from 3 to 10% of the strength of a covalent bond), the total energy released when a substrate is bound can be quite large, since it is usually bound at more than one site. Hydrogen-bonding and electrostatic interactions are the strongest of these intermolecular forces.

● Which of these forces could be effective in binding acetylcholine and noradrenaline to a receptor?

● The quaternary ammonium group of acetylcholine could be attracted to an anionic centre (like a carboxylate anion, $-COO^-$), by electrostatic forces; and the two oxygen atoms of the acetyl group could be at the 'receiving end' of hydrogen bonds. Noradrenaline could form hydrogen bonds through its three hydroxy and one primary amino groups.

The nature of the acetylcholine binding site at the receptor has been much investigated, but the exact geometry is still uncertain, and in fact there are known to be several different receptors, all of which possess a series of anionic binding sites (similar to those shown in Figure 1.1).

QUESTION 1.1

The functional groups on a hypothetical receptor are shown in Figure 1.2. What are the possible interactions between individual groups on the receptor and parts of the molecule $CH_3(CH_2)_8COCH_2CH_2NH_3^+$?

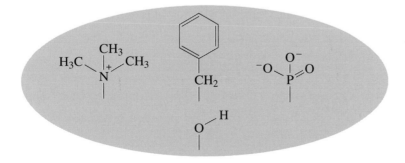

Figure 1.2 A hypothetical receptor surface.

* Intermolecular forces are discussed in Part 1 of *The Third Dimension*[1].

COMPOUNDS THAT MIMIC THE ACTION OF NORADRENALINE — AGONISTS

2

The use of drugs — chemical compounds that cause a physiological change in the human system — has a history as long as that of human civilization. (It was once suggested — not completely frivolously — that the difference between humans and other animals is the desire to take medicines to cure our ills.) In this Section we shall look at compounds that produce a similar physical response to noradrenaline when introduced into the body. These are said to be **agonists** (from the Greek *agonistes* meaning 'contestant') for noradrenaline, because they mimic its action.

At the end of the nineteenth and beginning of the twentieth century it was believed that adrenaline (**2.1**) was the principal neurotransmitter in the sympathetic nervous system. This belief was based on the fact that adrenaline had been isolated from the adrenal and pituitary glands of animals, and was found to induce the appropriate physiological response.

It was not until 1946 that it was discovered that noradrenaline was the principal neurotransmitter in the sympathetic nervous system, and that adrenaline acted as a noradrenaline agonist. Following that discovery, noradrenaline itself was marketed as a pressor (heart stimulating) agent for increasing blood pressure.

Another effective agonist for noradrenaline is ephedrine (**2.2**), which was discovered in 1760 BC by the pharmacist Emperor Shen Nung, who spent much of his life examining the potential of plant extracts for use as drugs.

In AD 1596 , a Chinese pharmacist ascribed to ephedrine the properties of improving circulation, causing sweating, easing coughing and reducing fevers — an excellent description of a noradrenaline agonist. In the 1920s, pure ephedrine was isolated, and by 1926 it was on general sale. Unlike synthetic adrenaline, ephedrine has the particular advantage that it is effective when taken orally. It became one of the most widely used drugs, finding special use in the treatment of asthma, where it helped prevent attacks of bronchospasm.

Ephedrine was originally derived from a Chinese plant, ma huang (Figure 2.1) — *Ephedra vulgaris* — which made it quite expensive. This stimulated research into finding cheap substitutes. After many years of research on ephedrine analogues, salbutamol (**2.3**), also known as Ventolin, was developed as a long-acting, highly potent drug for asthma relief. It is free of the excess heart stimulation or toxicity usually associated with other noradrenaline analogues. Despite having been in use for many years, Ventolin inhalers (Figure 2.2) are still the most common form of asthma treatment.

If we ask why compounds **2.1**, **2.2** and **2.3** mimic the action of the neurotransmitter noradrenaline, a reasonable hypothesis might be:

2.1 🖥
increases blood pressure

2.2 🖥
used in an early formulation for the treatment of asthma

2.3 🖥
market leader in the treatment of asthma

> If a compound has a structure that allows it to fit the receptor binding site in an almost exactly analogous way to the neurotransmitter, then it will be an agonist.

In other words, we suppose that there should be a *structure–activity relationship*.

● Look carefully at the structures of adrenaline (**2.1**), noradrenaline (**1.8**), ephedrine (**2.2**) and salbutamol (**2.3**). What is the common molecular fragment?

● They all contain a benzene ring, and a secondary amino group (−NHR) and a hydroxy (−OH) group on adjacent carbon atoms, as in the fragment **2.4**. So it looks as though binding could involve the hydrophobic benzene ring, and the polar −OH and −NH− groups.

2.4

If we label the carbon atom bearing the alcohol group as the 1-, or alpha (α-) carbon, and its neighbour the 2- or β-carbon atom, then such compounds are referred to as 2-aminoalcohols, or β-aminoalcohols.

There are other chemically interesting observations that we can make about the structures of noradrenaline and its agonists. Physiologically, the most highly stimulating of these compounds, with the longest-lasting effect, is ephedrine.

● How does ephedrine differ structurally from the others?

● It lacks *ortho* hydroxy or hydroxymethyl substituents on the benzene ring. Adrenaline and noradrenaline are derivatives of catechol (1,2-dihydroxybenzene), and salbutamol is a derivative of *ortho*-hydroxybenzylalcohol.

The absence of ring hydroxy or hydroxymethyl groups makes ephedrine a much less polar molecule than adrenaline, noradrenaline or salbutamol. Many biological membranes are constructed from *lipids* (commonly known as fats), water-insoluble molecules, whose structures are based on long-chain carboxylic acid derivatives of the trihydroxyalcohol, glycerol, and phosphoric acid esters, *phospholipids*. Less-

Figure 2.1
Growth of *Ephedra* spp. or ma huang.

Figure 2.2
Using a Ventolin inhaler.

polar molecules such as ephedrine (but not the other three) are more easily able to cross these relatively non-polar lipid* layers, which enables them to penetrate the brain. Once in the brain they bind to receptors that cause excitement and stimulation.

Although in this group of molecules, the naturally occurring ephedrine has the longest-lasting effect, *synthetic* drugs are often long-lasting because they do not react with the enzymes that specifically break down and remove *natural* neuro-transmitters. In salbutamol, for example, the $-CH_2OH$ group on the aromatic ring cannot be rapidly deactivated by the catechol-destroying enzymes that break down adrenaline and noradrenaline. The 1,1-dimethylethyl group, $(CH_3)_3C-$, on the nitrogen atom (commonly called the *tertiary*-butyl group) was also found to increase the affinity for binding in the lung receptors, while decreasing the affinity for the heart receptors. So, salbutamol has been designed to bind to noradrenaline receptors, not to cross into the brain because of its polarity, to have a long-lasting effect, and not to stimulate the heart — all this from a simple molecule containing 35 atoms, and only the elements C, H, N and O!

Interestingly, the noradrenaline agonists are structurally similar to another important, and familiar, class of drugs — the **beta-blockers** (Box 2.1).

* Recall the description of the structure of fats in Part 1, Box 1.4.

BOX 2.1 Professor Sir James Black

In 1959, James Black (Figure 2.3) at Imperial Chemical Industries formulated a theory for the possible treatment of coronary arterial disease, and in particular, angina. When the body is under physical or mental stress, the sympathetic nervous system is stimulated, causing an *increase* in the pulse rate of the heart and its force of contraction. These processes require energy, which means that the heart muscles require a reliable supply of oxygen. If the supply of oxygen from the blood is restricted because of coronary arterial disease, intense angina pain is experienced. Black proposed that the heart's demand for oxygen could be reduced by blocking the effects of stimulating the sympathetic nervous system; that is, by preventing the rise in heart rate and contracting force. Because the link between the stimulation of the sympathetic nervous system and the increase in heart rate is mediated by receptors known as beta-receptors, compounds that can treat angina

in this manner are known as *beta-blockers*. These compounds also relieve the stress associated with examinations, public speaking, and interviews (for example), and their use and abuse in sport has been a subject of much dis-cussion. Two commercially successful beta-blockers are practolol (**2.5**) and propranolol (**2.6**).

Figure 2.3
Sir James Black.

CH₃CONH

2.5 🖥
used in the treatment of angina and hypertension

2.6 🖥
used in the treatment of angina

However, it is not just the nature of the functional groups, or where they are on the carbon skeleton that is important. So far, we have not alluded to the one remaining structural feature common to most of compounds **2.1–2.6**.

● Synthetic adrenaline was found to be only half as active as natural adrenaline. Why might that be?

● Adrenaline has a chiral centre, and therefore the natural product will be only one of a pair of enantiomers. If no provision were made in the synthesis to generate or isolate only one of the enantiomers, the synthetic product would be a racemic mixture — that is, equal quantities of the two enantiomers.

COMPUTER ACTIVITY 2.1
Stereochemistry of noradrenaline agonists

In this activity (using WebLab ViewerLite images on the CD-ROM associated with this Book) you will first compare the stereochemistry of noradrenaline with adrenaline, ephedrine and benzedrine, and then with salbutamol, practolol and propranolol.

It should take you about 15 minutes to complete this activity.

Open WebLab ViewerLite.

Using the 'Open' command from the 'File' menu, load from the CD-ROM the models of adrenaline (WebLab file CE1), noradrenaline (WebLab file CE2), ephedrine (WebLab file CE3) and benzedrine (WebLab file CE4), into separate screens. If the models are not already in the 'Ball and Stick' style, convert them all to this form.

Display these four screens simultaneously using the 'Tile horizontal' command from the 'Windows' menu.

Notice that, where appropriate, in each molecular model the hydroxyl group and the amino group nitrogen are eclipsed, and the oxygen, nitrogen and the two central carbon atoms, are all in the plane of the screen.

● What do you notice about the stereochemistry of the chiral carbon atom bearing the hydroxyl group (carbon-1) in adrenaline (CE1), noradrenaline (CE2) and ephedrine (CE3)?

● All three of the compounds have the same configuration at the C-1 carbon atom.

Notice that benzedrine (CE4) does not have an OH group on C-1, so the stereochemistry cannot be compared.

Now retain the noradrenaline screen (CE2), but close the other three screens.

Load salbutamol (CE5), practolol (CE6) and propranolol (CE7) into three new WebLab ViewerLite screens.

Again choose 'Tile horizontal' from the 'windows' menu and compare the configuration at the hydroxyl-bearing carbon atom, C-1, of the three new models with noradrenaline.

You should find that these three compounds also have the same configuration at the chiral carbon atom bearing the hydroxyl group (carbon-1)

2.1 Summary of Sections 1 and 2

1   In the nervous system, muscles are stimulated by chemical mediators called neurotransmitters. Acetylcholine is the neurotransmitter in the parasympathetic nervous system, and both acetylcholine and noradrenaline act in this way in the sympathetic nervous system.

2   An agonist is a compound that mimics the action of a naturally occurring, physiologically active molecule when introduced into the body as a drug. Adrenaline and ephedrine are agonists for the neurotransmitter noradrenaline.

3   Neurotransmitters bind to receptors through intermolecular forces — hydrogen-bonding, electrostatic interactions, London forces and hydrophobic interaction.

4   A neurotransmitter is usually bound to at least two sites on a receptor, which accounts for its specific activity.

5   The —C(OH)—C(NHR)— group is a common feature in many noradrenaline agonists and beta-blockers.

QUESTION 2.1

Is the configuration at carbon-1 in the noradrenaline agonists and the beta-blockers R- or S-? Indicate how you reached your decision.

QUESTION 2.2

Look at the structures of mescaline (**2.7**, Figure 2.4), a hallucinogenic drug originally used by Mexican Indians, rimiterol (**2.8**) and naphazoline (**2.9**). Which of these do you expect to be noradrenaline agonists? Which contain chiral centres? How many stereoisomers are there of each drug? How are the isomers related stereochemically?

2.7   **2.8**   **2.9**

Figure 2.4
Cactus from which mescaline is obtained.

THE TARGET: β-AMINOALCOHOLS

3

It is important to recognize that almost all drugs, including most of compounds **2.1–2.6**, are synthetic materials. They cannot be obtained from natural sources, or not in sufficient quantity, since some drugs are manufactured on the tonne scale. In this Section, we shall consider the synthesis of pseudoephedrine, an active ingredient in medicines such as Actifed® or Sudafed® (note the corruption of the chemical name in these trade names!), which are used to relieve nasal and respiratory congestion (Figure 3.1). About 30 mg of the compound are contained in each 5 cm³ spoonful.

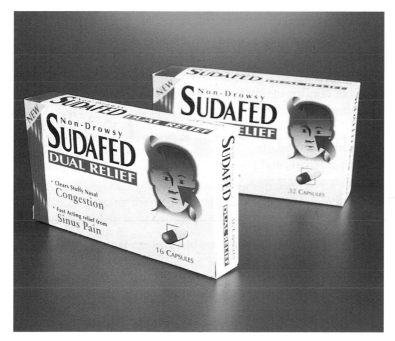 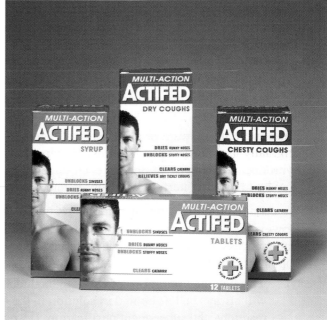

Figure 3.1 Medicinal preparations containing pseudoephedrine.

Pseudoephedrine has an action similar to that of ephedrine, but has less pressor (heart-stimulating) action. It is present in extracts of the plants *Ephedra* spp., but it cannot be obtained commercially from this source.

Although we are concentrating on just one example, all the compounds **2.1–2.6** have quite similar structures, and we shall see later that the strategy used to synthesise pseudoephedrine can be extended to these other compounds.

The goal of synthetic chemists is to construct the compounds they want to make — their **target molecules** — from readily available starting materials. This can rarely be achieved by a single step, and usually the molecule has to be built up using a series of reactions. This collection of stages, the pathway between the starting materials and the target molecule, is referred to as the **synthetic route**. The term **'readily available starting material'**, has a rather flexible usage, but a useful working definition would be 'a chemical obtainable from a commercial supplier at a cost within the laboratory budget'!

5.5 How efficient was the synthesis?

The synthesis of the racemic mixture, and the percentage yields obtained, are shown in Scheme 5.2:

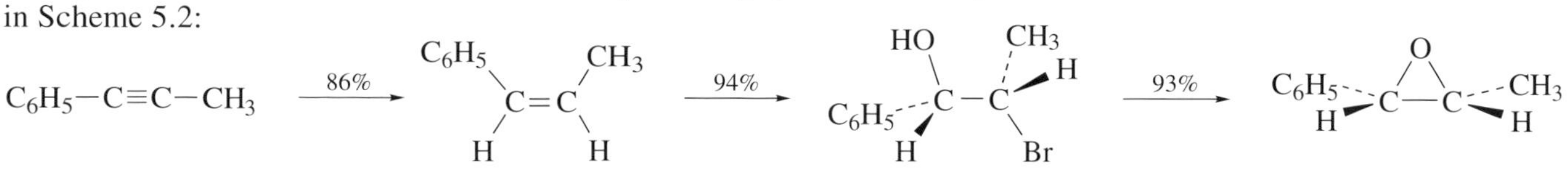

plus its enantiomer   plus its enantiomer

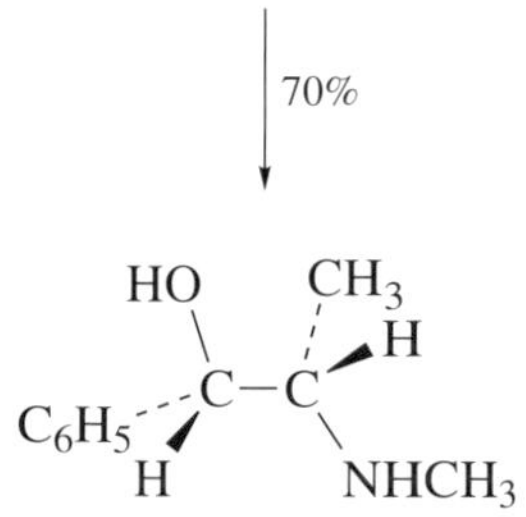

Notice that in going from one stage to the next, some material is lost: 14% in the first stage, 6% in the next, 7% in the next, and so on. To get an idea of the overall efficiency of the route when all the stages are taken together, we multiply the percentage yields of each individual stage (expressed as fractions), and express the result as a percentage. In this case, the **overall yield** of the racemic mixture is:

$$\frac{86}{100} \times \frac{94}{100} \times \frac{93}{100} \times \frac{70}{100} = 53\%$$

This doesn't seem very good. Almost half of the original material is lost in just getting to the racemic mixture; obtaining the single enantiomer wastes at least another 50% of the product automatically. In reality, however, for a four-stage route this is not such a bad result. In a multi-stage synthesis, the yields for each individual stage need to be high if the overall yield is to be reasonable. This is emphasized in the data in Table 5.3; the overall yield decreases dramatically as the number of stages increases, or as the yield per stage decreases.

plus its enantiomer

SCHEME 5.2

Table 5.3 The percentage overall yield as a function of the number of stages and the average yield per stage

Number of stages	Average yield per stage				
	90%	80%	70%	60%	50%
2	81	64	49	36	25
3	73	51	34	22	13
4	66	41	24	13	6
6	53	26	12	5	1.6
10	35	11	3	0.6	0.10

It is important that you realize that really high-yielding reactions are needed for an efficient synthetic route.

5.6 Summary of Section 5

1   The amounts of reactants required in a reaction are calculated on a molar basis.

2   The choice of solvent depends both on the solubility of the reactants and mechanistic considerations.

3   The percentage yield is given by:

$$\text{percentage yield} = \frac{\text{mass obtained}}{\text{theoretical yield}} \times 100$$

The theoretical yield is that expected assuming 100% transformation of the number of moles of the starting material that is used in the lowest molar amount.

4   Racemic mixtures can be separated by reaction with a homochiral reagent. This is known as resolution.

5   The overall yield is calculated by multiplying together the percentage yields of each stage (expressed as fractions), and expressing the result as a percentage.

QUESTION 5.1

In Reaction 5.7 we started with 22.0 g of the chiral aziridine **5.21**, and the yield of product **5.22** was 21.6 g. What was the percentage yield? Begin by working out the molar mass of the starting material and the product. (Alternatively, you can use the % molar mass calculator on the CD-ROM to answer this question.)

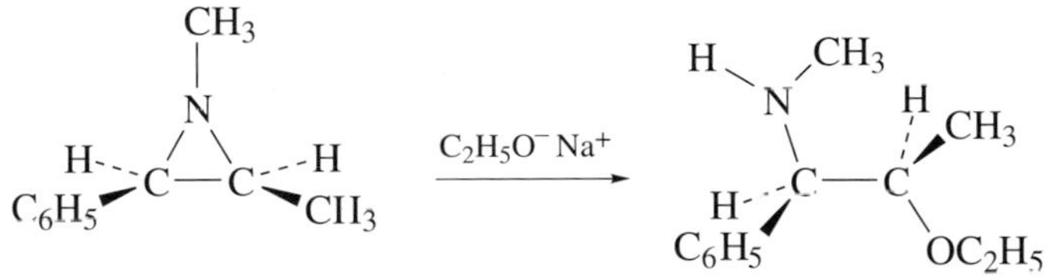

(5.7)

QUESTION 5.2

Use the molar percentage yields for the two separate steps in Scheme 5.2 (1-phenylprop-1-ene to the bromoalcohol mixture, and bromoalcohols to oxiranes) to calculate an overall percentage yield for the direct conversion of alkene to oxiranes.

QUESTION 5.3

The conversion of **5.21** into **5.22** was the last stage of a four-stage synthesis, shown schematically below. Using your answer to Question 5.1, work out the overall percentage yield of **5.22**.

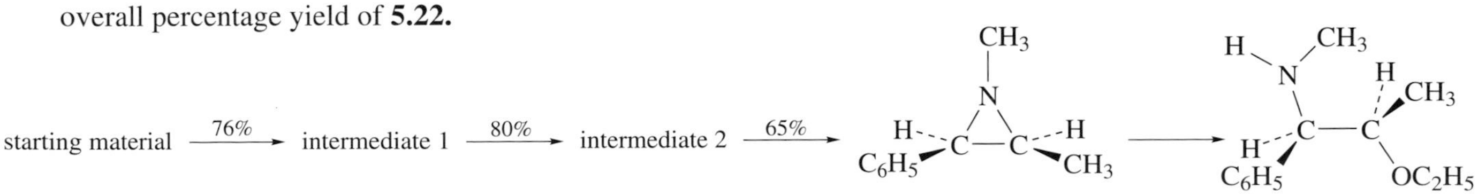

(5.8)

THE SYNTHESIS OF β-AMINOALCOHOLS

6

One way in which new drugs are developed involves making analogues of a known, usually natural, compound that has some biological activity, and then testing these analogues to discover if they are more active or cause fewer side-effects. By analogues, we mean compounds that retain the structural features and active functions of the original compound, but whose structure is slightly modified.

Let's see how this approach works with pseudoephedrine.

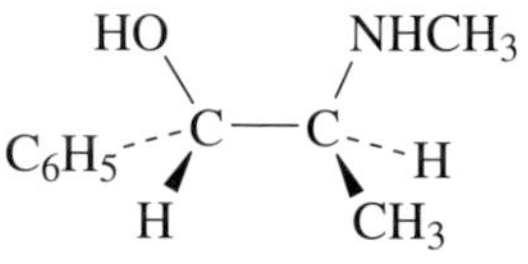

● How do you think pseudoephedrine (**4.1**) binds to a receptor?

● Hydrogen-bonding may be important for the $NHCH_3$ and OH groups, and the phenyl group may bind through hydrophobic interactions. The $NHCH_3$ group may be protonated to $^+NHCH_3$, which could be involved in electrostatic interactions.

4.1

So, if we want to change the structure yet maintain binding at the receptor, we need to keep these features fairly constant.

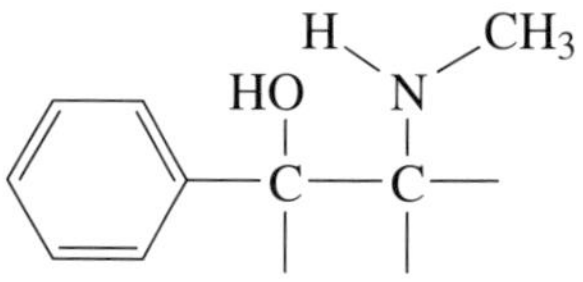

● Suggest some possible analogues of pseudoephedrine.

● Keeping the structural feature **6.1** constant, some possibilities are **6.2–6.4**, in which R is an alkyl group (you may have come up with some good alternatives):

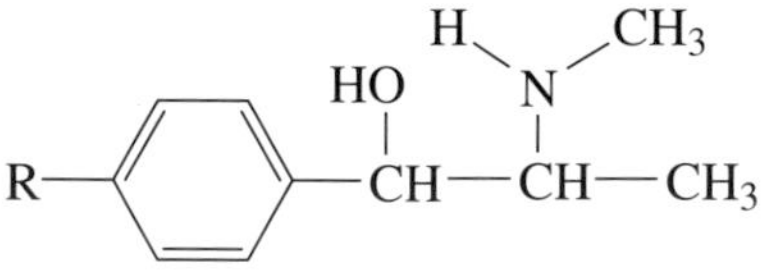

6.1

6.2

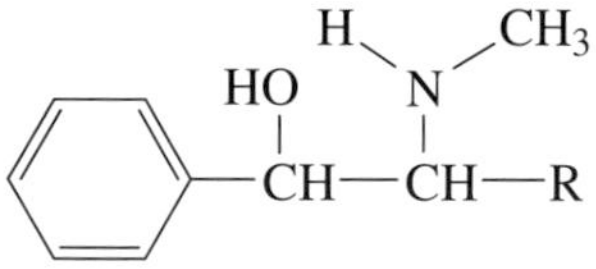

6.3

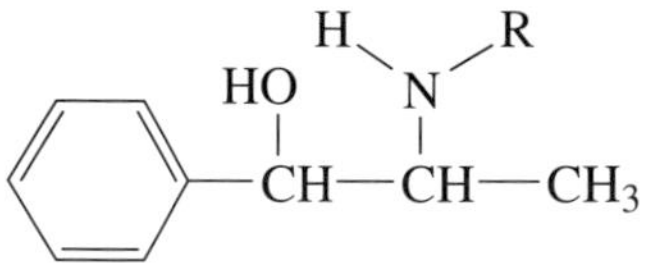

6.4

● Using our synthesis, how could we make **6.2–6.4**?

● Compounds **6.2** and **6.3** could be made using the same synthetic route, but starting with different alkynes, **6.5** and **6.6**, respectively. Compounds **6.4** can be made by using a different alkylamine (RNH_2) in the last step.

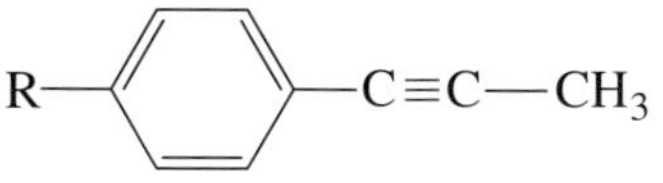

6.5

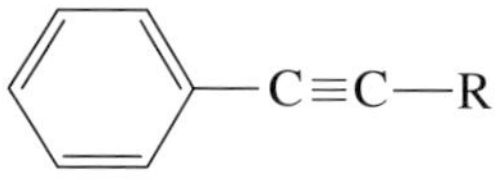

6.6

Once a synthetic route has been developed, different reagents can be chosen, and a range of similar compounds can be generated for testing. This is the advantage of designing a flexible synthesis: we don't have to devise a completely separate route for each analogue.

An ingenious modern way of doing this is called *combinatorial chemistry* (see Section 1). If we were able to obtain ten different alkynes and ten different amines, for example, this would generate a total of $10 \times 10 = 100$ possible analogues. Instead of working through a complete synthesis for every different compound on a similar scale to our synthesis of pseudoephedrine, an automatic synthetic procedure is devised, working on a very small scale. In a short time, tiny quantities of one hundred new compounds are ready for testing. Since rapid automated methods are also available for assaying the activity of new compounds, all hundred products could easily be subjected to preliminary testing, even before separating the stereo-isomers. Any compound showing promise could then be synthesised in greater quantities for more thorough investigation.

QUESTION 6.1

How would you synthesise propranolol (**2.6**), which is used in the treatment of angina? How would the procedure differ from that needed for the synthesis of pseudoephedrine?

2.6

A DIFFERENT WAY OF LOOKING AT SYNTHETIC EFFICIENCY

7

7.1 'Green chemistry'

Look again for a moment at the procedure we outlined in Section 5.3 for the conversion of Z-1-phenylprop-1-ene into the mixture of bromoalcohols (Scheme 7.1).

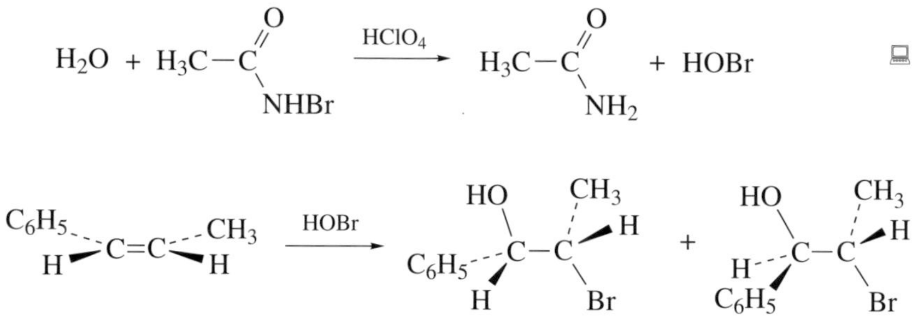

SCHEME 7.1

The percentage yield for this reaction, 94%, was reasonably good, but does this figure really indicate the true efficiency of this reaction step? We must ask ourselves a couple of other questions first.

● We lost 6% of our starting material, but what else might we have thrown away during the course of this reaction?

● You should have noted the aqueous perchloric acid ($HClO_4$), and the ethanamide (CH_3CONH_2), the acid catalyst and the byproduct from the first reaction. Remember too that the N-bromoethanamide was used in excess. Perhaps you thought too about the reaction solvents that would be needed, but maybe not about any materials required in the isolation and purification of the products. In addition, we would use water, ice for cooling, and energy to encourage the reaction to go and to evaporate solvents.

● For example, how much of the 34.5 g (0.25 mol) of N-bromoethanamide eventually ended up in our product?

● Only 60% of the reagent could be used at most, as we used 0.25 mol when the stoichiometric amount was 0.15 mol (20.7 g). Then, only 0.15 mol at most of the bromine (12.0 g) is incorporated into the bromoalcohol; the rest of the material (34.5 g − 12.0 g = 22.5 g) joins the solvents and reagents down the waste pipe.

It was considerations like these that led a number of synthetic chemists, particularly in the chemical industry, to re-evaluate the principle that a high percentage yield was an appropriate efficiency goal. In the United States alone, more than ten million tonnes of hazardous chemicals in 1997 were treated, recycled, used for energy production, disposed of, or released into the environment. The costs of waste disposal have increased with every new piece of legislation that sought to protect the environment from the degradation caused by the uncontrolled dumping of chemical wastes, to the point where for many chemical companies, the cost of dealing with environmental regulations exceeds their expenditure on research. It has been said that 'what one does not produce from a chemical reaction is almost as important as what one does produce!'

Out of this situation, the **green chemistry movement** was born. Its advocates suggested that chemical production should be viewed in a new way. It was not sufficient to be able to control, treat, clean up and dispose of hazardous waste.

> The aim should be not to generate it in the first place!

This is the first point on '*The twelve principles of green chemistry*', the charter for the movement (see Box 7.1, p.128). Other principles that are particularly relevant for synthesis include:

- Incorporate a large part of each reagent used in the synthetic steps into the final product.
- Reduce the use of auxiliary substances, like solvents — for reaction, extraction and purification — and drying agents.
- Use catalytic reagents where possible in place of stoichiometric reactants.
- Utilize renewable raw materials where possible.
- Minimize use of energy for heating, cooling, mixing, pressure and transport.

One particular concept to arise from this new perspective is that of **atom economy**. Instead of concentrating solely on the main starting material and the required product, atom economy considers every reagent used in the synthetic step. We can demonstrate this approach, by returning again to the hydroxybromination of Z-1-phenylprop-1-ene. To quantify the atom economy of a reaction, we use the following formula:

$$\% \text{ atom economy} = \frac{\text{formula mass of the required product}}{\text{sum of the formula masses of all the reactants used in the reaction}} \times 100\%$$

Formula mass is another term for the mass of one mole (expressed in grams). We can now apply the formula* to the addition reaction that yields the bromoalcohols in the pseudoephedrine synthesis.

$$C_6H_5CH{=}CHCH_3 + CH_3CONHBr + HOH \longrightarrow C_6H_5CHOHCHBrCH_3 + CH_3CONH_2 \quad (7.1)$$

C_9H_{10}	$C_2H_4NO + Br$	$H + OH$	$C_9H_{11}OBr$	C_2H_5NO
118 g	58 g	80 g 1 g 17 g	215 g	59 g

The *formula mass of the required product* is the formula mass of $C_9H_{11}OBr$, 215 g (printed in *green*); the sum of the *formula masses of all the reactants used in the reaction* is 274 g. The actual atoms of the reactants utilized in constructing the product, and their corresponding formula masses (118 + 80 + 17) g are shown in *green*, and those not utilized (58 + 1) g are shown in *red*. The % atom economy is therefore 215/274 × 100 = 78%. Even this reckoning does not include the reagents used in excess, and any allowance for the solvents, auxiliary chemicals or energy.

---

* In calculating the sum of the formula masses of the reactants you have to take account of the stoichiometry of the reaction; for example, if two molecules of water are involved, the appropriate formula mass of water to use is 36 g not 18 g.

7.2 'Green chemistry' in action

Ibuprofen* (**7.1**) was discovered at the Boots Chemical Company in Nottingham, who patented a synthesis for its manufacture on a large scale in the 1960s. It is an analgesic (pain killer), anti-inflammatory, and antipyretic (fever-reducing) drug, with a better specification and fewer side-effects than aspirin or paracetamol. It is probably the most effective painkiller licensed for over-the-counter sale to the general public.

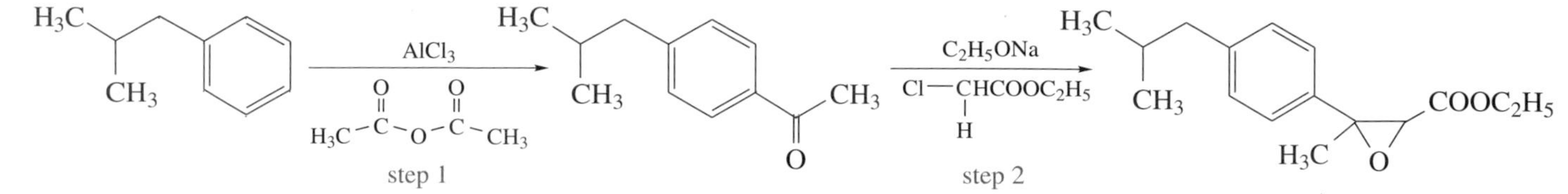

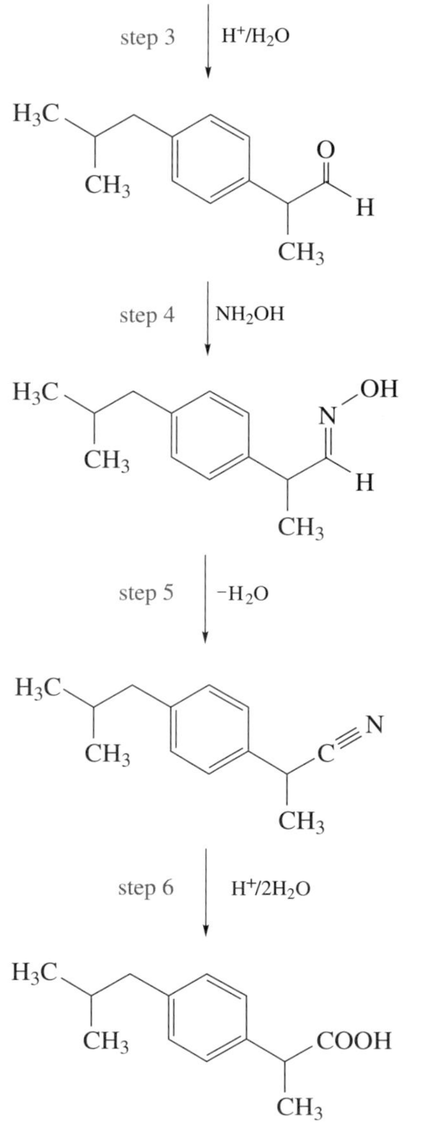

The Boots synthesis is shown in Scheme 7.2. It was designed to be economical and industrially manageable, but not to be environmentally friendly. This synthesis was employed to produce about fourteen thousand tonnes of ibuprofen annually, but unfortunately its use also generates more than sixteen thousand tonnes of unwanted byproducts or waste! Some of the reactions in this Scheme, and the reagent hydroxylamine, NH_2OH, may be unfamiliar to you, but do not worry about that.

In Scheme 7.2 the reactant atoms shown in red are the ones that are eventually incorporated into the final product, but it is salutary to note how many atoms are thrown away. Table 7.1 also shows this in a more quantitative way, which enables us to calculate the percentage atom economy of the synthetic route.

Note that we do not include the H^+ in the calculation, as it is a catalyst. Nevertheless, a considerable quantity of acid is required to bring about the reactions in steps 3 and 6.

We can now do the calculation:

$$\% \text{ atom economy} = \frac{206}{647} \times 100 = 32\%$$

This is not a very satisfactory result, especially when you consider that the waste products include metallic (step 2) and ammonium salts (step 6), ethanoic acids (step 1).

Once the patents on ibuprofen had expired, the BASF–Hoechst–Celanese consortium (BHC) sought to develop a new synthesis, which was more atom-economical, but still a financially viable economic process. No company is going to embrace a green policy that significantly reduces profits! The manufacturing process that they were eventually able to bring on-stream in 1992 is shown in Scheme 7.3. Raney nickel is a specially prepared form of nickel with a very large surface area, which makes it an excellent (but not too expensive) catalyst for hydrogenation reactions. Palladium is also a commonly used catalyst in industry, despite its high cost. The BHC Consortium operates one of the largest ibuprofen plants in the world, annually producing enough ibuprofen for six billion tablets, well over three thousand tonnes.

7.1 🖥

SCHEME 7.2

* Ibuprofen synthesis is also discussed in Part 4 of *Mechanism and Synthesis*[4].

Table 7.1 Atom economy in the Boots synthesis of ibuprofen

Step	Reactant Formula	FM/g*	Utilized in ibuprofen Formula†	FM/g	Not utilized in ibuprofen Formula*	FM/g
1	$C_{10}H_{14}$	134	$C_{10}H_{13}$	133	H	1
1	$AlCl_3$	133.5			$AlCl_3$	133.5
1	$C_4H_6O_3$	102	C_2H_3	27	$C_2H_3O_3$	75
2	$C_4H_7ClO_2$	122.5	CH	13	$C_3H_6ClO_2$	109.5
2	C_2H_5ONa	68			C_2H_5ONa	68
3	H_2O	18			H_2O	18
4	NH_3O	33			NH_3O	33
6	H_4O_2	36	HO_2	33	H_3	3
Total	$C_{20}H_{42}NO_{10}Cl_4AlNa$	647	$C_{13}H_{18}O_2$	206	$C_7H_{24}NO_8Cl_4AlNa$	441

* FM = formula mass.

† The word 'formula' here does not refer to the molecular formula of individual compounds, but to the atoms that are or are not utilized in the ibuprofen product.

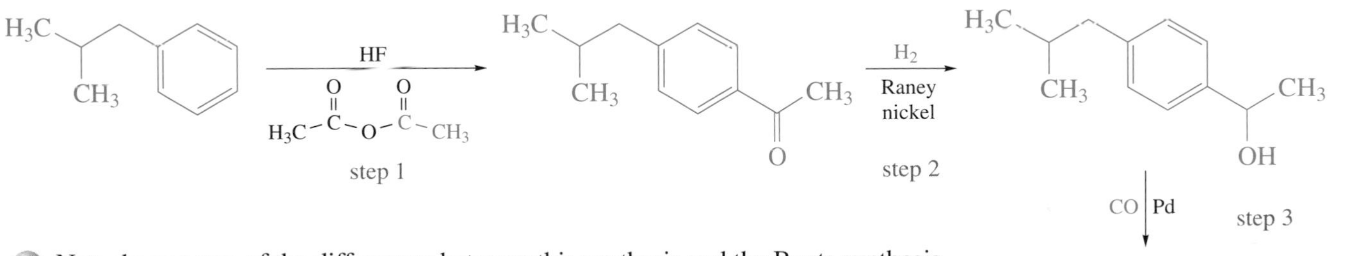

SCHEME 7.3

● Note down some of the differences between this synthesis and the Boots synthesis discussed earlier. Which of the 'principles of green chemistry' does it employ?

● The most obvious difference is that ibuprofen is now produced in a three-stage process in comparison with the six-stage Boots process. This allows for greater throughput, with a saving in both manufacturing time and capital expenditure. Thus, the green principles of low energy use, and reduced need for auxiliary substances are observed. The other green features are the repeated use of catalysts (Raney nickel, palladium, and hydrogen fluoride) rather than stoichiometric procedures, and the high level of incorporation of reagent atoms into the final product.

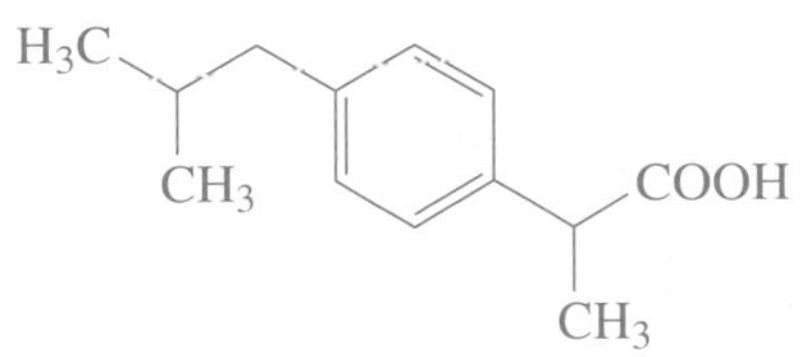

Just how effectively this last feature has been achieved is seen by carrying out an atom economy calculation (Table 7.2, p. 129). The catalysts, Raney nickel and palladium, are both recovered for further use, and even the hydrogen fluoride (which, unlike aluminium chloride, is used in only catalytic quantities) can be recycled.

BOX 7.1 The twelve principles of green chemistry

(after Paul T. Anastas and John Warner)

1 Prevention is better than cure.

It is better to prevent waste than to treat or clean up waste after it is formed.

2 Incorporate all materials used in the synthesis into the final product.

Synthetic methods should be designed to maximize the incorporation of all materials used in the process into the final product.

3 Ensure low risk to human health and the environment.

Wherever practicable, synthetic methodologies should be designed to use and generate substances that possess little or no toxicity to human health and the environment.

4 Design for efficacy of function with reduced toxicity.

Chemical products should be designed to preserve efficacy of function while reducing toxicity.

5 Eliminate or reduce auxiliary substances.

The use of auxiliary substances (e.g. solvents, separation agents, etc.) should be made unnecessary wherever possible and, innocuous when used.

6 Minimize energy requirements.

Energy requirements should be recognized for their environmental and economic impacts, and should be minimized. Synthetic methods should be conducted at ambient temperature and pressure.

7 Choose renewable raw materials.

A raw material or feedstock should be renewable rather than depleting, wherever technically and economically practicable.

8 Avoid unnecessary derivatization.

Unnecessary derivatization (blocking group, protection/deprotection, temporary modification of physical/chemical properties) should be avoided whenever possible.

9 Use catalytic reagents not stoichiometric reagents.

Catalytic reagents (as selective as possible) are superior to stoichiometric reagents.

10 Chemical products should not persist in the environment.

Chemical products should be designed so that at the end of their function they do not persist in the environment, but break down into innocuous degradation products.

11 Develop analytical methodologies for real-time in-process monitoring.

Analytical methodologies need to be further developed to allow for real-time in-process monitoring and control prior to the formation of hazardous substances.

12 Minimize the potential for chemical accidents.

Substances and the form of a substance used in a chemical process should be chosen so as to minimize the potential for chemical accidents, including releases, explosions and fires.

⬤ What is the percentage atom economy of this process?

⬤ % atom economy = 206/266 × 100 = 77%

This is a significantly better use of reagents and starting materials than the Boots synthesis, but still generates healthy financial returns in a situation where ibuprofen manufacture is an increasingly competitive business. In recognition of their success at 'greening' a large-scale industrial process, BHC won the 'Alternative Synthesis' section of the prestigious US Presidential Green Chemistry Challenge Award in 1997.

Table 7.2 Atom economy in the BHC synthesis of ibuprofen

Step	Reactant Formula* FM/g†		Utilized in ibuprofen Formula FM/g		Not utilized in ibuprofen Formula FM/g	
1	$C_{10}H_{14}$	134	$C_{10}H_{13}$	133	H	1
1	$C_4H_6O_3$	102	C_2H_3O	43	$C_2H_3O_2$	59
2	H_2	2	H_2	2		
3	CO	28	CO	28		
Total	$C_{15}H_{22}O_4$	266	$C_{13}H_{18}O_2$	206	$C_2H_4O_2$	60

* The word 'formula' here does not refer to the molecular formula of individual compounds, but to the atoms that are or are not utilized in the ibuprofen product.

QUESTION 7.1

Which of the three basic reaction types — substitution, addition and elimination — has the highest potential for use in green syntheses? What features reduce the usefulness ('greenness') of the other two?

QUESTION 7.2

Calculate and compare the % atom economy of the two routes shown for the synthesis of methyl 2-methylprop-2-enoate (methyl methacrylate), **7.2**, the monomer for the production of the polymer 'Perspex'. (Don't worry about the chemistry involved.) Note that the H_2SO_4 in route 1 and the palladium (Pd) in route 2 are catalysts, and should not be included in the % atom economy calculations. You can use the spreadsheet to do this.

7.3 Summary of Sections 6 and 7

1   A synthetic route can be used to make a variety of similar compounds by changing the reagents used in the various stages.

2   Twelve principles of green chemistry have been proposed, encouraging the designing of syntheses for the reduction of waste.

3   % Atom economy is defined as

$$\% \text{ atom economy} = \frac{\text{formula mass of the required product}}{\text{sum of the formula masses of all the reactants used in the reaction}} \times 100\%$$

LEARNING OUTCOMES

Now that you have completed *Alkenes and Aromatics: Part 3 — A first look at synthesis*, you should be able to do the following things:

1   Recognize valid definitions of, and use in a correct context, the terms, concepts and principles in the following Table. (All Questions)

List of scientific terms, concepts and principles introduced in *A first look at synthesis*

Term	Page number	Term	Page number
agonist	96	overall yield	120
atom economy	125	percentage yield	113
Avogadro's hypothesis	112	readily available starting material	101
beta-blocker	98		
combinatorial chemistry	93	resolution (of enantiomers)	118
crude yield	113	stoichiometric amount	115
formula mass	125	structure–activity relationship	93
fractional crystallization	119	synthesis	91
green chemistry movement	125	synthetic route	101
homochiral	119	target molecule	101
molar scale	111	theoretical yield	113
neurotransmitter	94		

2   Explain how a molecule can bind to a receptor. (Question 1.1)

3   Identify possible noradrenaline agonists and beta-blockers. (Question 2.2)

4   Apply the Cahn–Ingold–Prelog rules for specifying stereochemistry at a chiral centre to naturally occurring molecules and synthetic drugs. (Questions 2.1, 4.1 and 4.2)

5   Devise simple organic synthetic sequences that involve only functional group changes. (Questions 4.3, 4.4, 4.5 and 6.1)

6   Modify existing synthetic routes, to give new target compounds. (Questions 4.4 and 6.1)

7   Calculate the percentage yield of the individual stages, and the overall yield, of a reaction sequence. (Questions 5.1, 5.2 and 5.3)

8   Outline some of the principles of green chemistry, and show how they may be applied. (Question 7.1)

9   Calculate % atom economy for a reaction. (Question 7.2)

QUESTIONS: ANSWERS AND COMMENTS

QUESTION 1.1 (*Learning Outcome 2*)

First look for particular functional groups in $CH_3(CH_2)_8COCH_2CH_2NH_3^+$. The molecule can be split into imaginary component parts:

$$CH_3CH_2CH_2CH_2CH_2CH_2CH_2CH_2CH_2\overset{\displaystyle O}{\overset{\displaystyle \|}{-C}}-CH_2CH_2-NH_3^+$$

hydrocarbon   carbonyl   ammonium
chain   group   group

The hydrocarbon tail is most likely to take part in hydrophobic or London interactions. The hydrophobic, or 'water-hating,' part of a molecule is the non-polar region consisting only of carbon and hydrogen atoms. The most likely site on the receptor for such interaction is the benzene ring.

The carbonyl group is polarized; the oxygen atom is susceptible to hydrogen-bonding with acid protons, but is not sufficiently polarized to form a significant electrostatic interaction with charged sites on the receptor. The carbonyl group could therefore hydrogen-bond with the hydroxy group on the receptor.

The ammonium group bears a full positive charge, and is therefore likely to form a significant electrostatic interaction with the phosphate group, $-PO_3^{2-}$, on the receptor.

QUESTION 2.1 (*Learning Outcome 4*)

Although the configuration at C-1 in adrenaline, ephedrine, and salbutamol is *R*-, that of the beta-blockers is *S*-. The orientation of the $-OH$ group attached to C-1, the aromatic side-chain, C-2 bearing the amino group, and the hydrogen atom attached to C-1 remains the same, but the group priority changes from

$$-OH > -CH_2NHR > aromatic > -H$$

in the agonists, to

$$-OH > -CH_2O-aromatic > -CH_2NHR > -H$$

in the beta-blockers. The principle employed here is the Cahn–Ingold–Prelog rule that carbon attached to oxygen has higher priority than carbon attached to nitrogen.

ephedrine (*R*-)   propanolol (*S*-)

Q.1   **Q.2**

In fact, only *R*-adrenaline, and *R*-noradrenaline are effective; the enantiomers **Q.3** and **Q.4** are inactive. So although the structural fragments look simple, the receptor site is certainly chiral and rather complex.

S-noradrenaline (R = H; **Q.3**)
S-adrenaline (R = CH_3; **Q.4**)

QUESTION 2.2 (Learning Outcome 3)

Mescaline, rimiterol (an asthma-controlling drug) and naphazoline (a vasodilator, used as a nasal decongestant) are all, in fact, noradrenaline agonists. Although only rimiterol contains fragment **2.4** (slightly disguised), they all contain hidden in their structures the fragment shown as **Q.5**. This suggests that compounds with only two of the three active groups are also able to interact with the noradrenaline receptors.

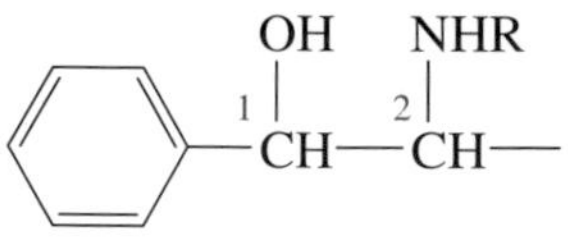

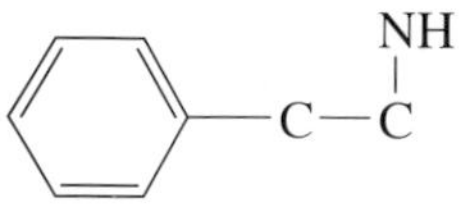

Mescaline and naphazoline have no chiral centres, so exist only as a single stereoisomer. Rimiterol has two chiral centres, so there are four (2^2) stereoisomers:

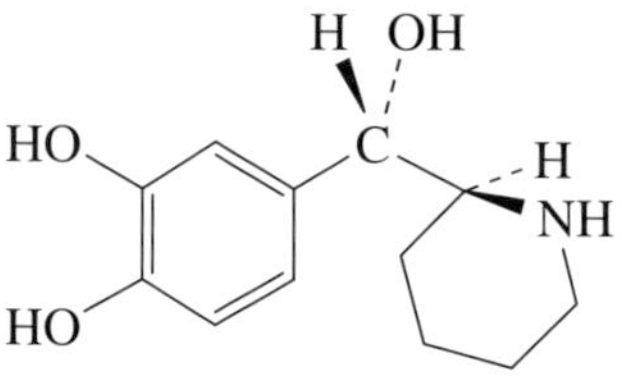

The pairs of enantiomers are (a) and (b), and (c) and (d). Pairs of diastereomers are (a) and (c), (a) and (d), (b) and (c), and (b) and (d).

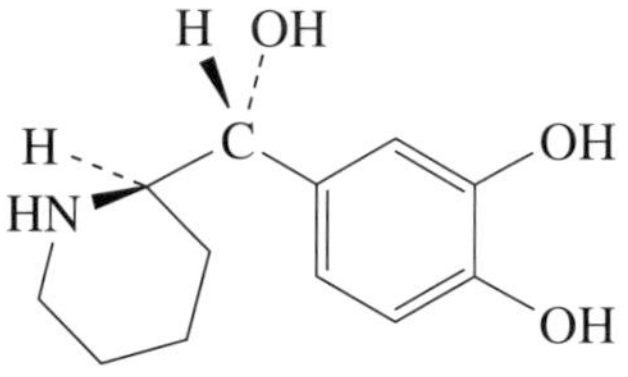

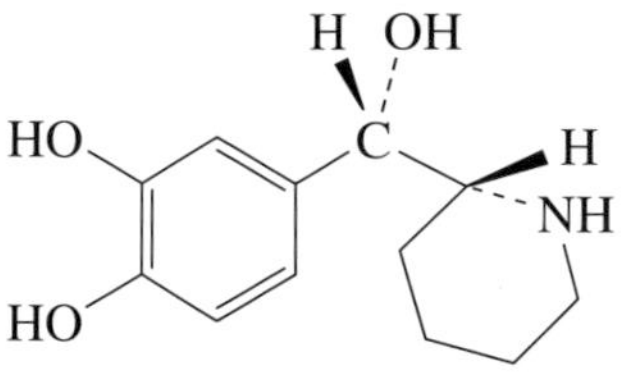

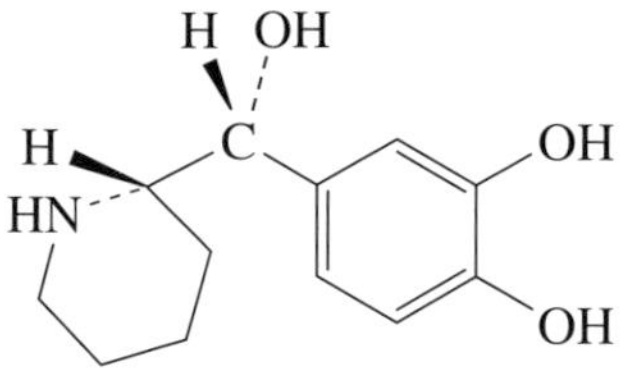

QUESTION 4.1 (Learning Outcome 4)

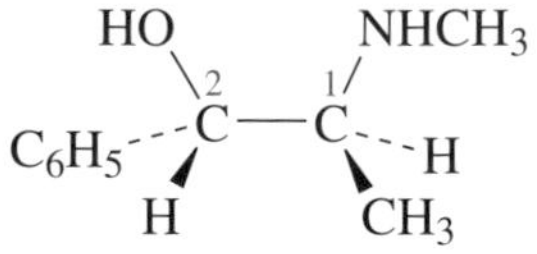

First rotate the left-hand chiral centre in **4.1** until the hydrogen atom is pointing away from you. The priority order of groups at this carbon is $-OH > C-NHCH_3 > C_6H_5$. As these groups are arrayed anticlockwise, the stereochemistry at this centre (carbon-1) is S-. The priority order at carbon-2 is $NHCH_3 > C-OH > CH_3$, which are also arrayed anticlockwise, so this centre has S-stereochemistry too. The complete systematic name of pseudoephedrine is therefore 1S,2S-2-methylamino-1-phenylpropan-1-ol.

4.1

QUESTION 4.2 (*Learning Outcome 4*)

In 1-methyl-2-phenyloxirane the priorities at C-1 are $O > CH(C_6H_5)O$ (carbon attached to oxygen) $> CH_3$ (carbon attached to hydrogen), so this centre is designated $1R$-. At C-2, the priorities are $O > CH(CH_3)O > C_6H_5$ (carbon attached to carbon), so this centre is designated $2S$-.

Since oxirane **4.9** has an R configuration at carbon atom 1 and an S configuration at carbon atom 2, it is known as $1R,2S$-1-methyl-2-phenyloxirane. Despite such complications, the important message at this stage is that, by considering the mechanism of our chosen reaction, we have been able to predict sensibly which reactant is required to get to pseudoephedrine.

QUESTION 4.3 (*Learning Outcome 5*)

The first step, the Friedel–Crafts acylation, is expected to go well; there will be no rearrangement of the side-chain.

However, in step 3, the elimination of water from an alcohol using the dehydrating agent concentrated sulfuric acid, produces mainly the *trans* (E-) isomer. So although the conjugation in the product will ensure that 1-phenylprop-1-ene and not 1-phenyl-prop-2-ene is formed, the stereochemistry about the carbon–carbon double bond will not be suitable for our synthetic route.

QUESTION 4.4 (*Learning Outcomes 5 and 6*)

As with any synthesis, we work backwards, focusing on the functional groups. By analogy with pseudoephedrine, the $-C(OH)-C(NHR)-$ arrangement of groups can be made from an oxirane. We saw earlier that the *trans*-1-methyl-2-phenyloxirane (**4.7**) reacts with methylamine to give ephedrine:

(Q.1)

The oxirane results from the action of a base on the corresponding bromoalcohol, which is made by *anti* addition of HOBr to E-1-phenylprop-1-ene:

(Q.2)

Because the E-alkene is commercially available, this is our starting point. Of course, a racemic mixture of bromoalcohols is generated in the first reaction, so the two enantiomers need to be separated at some stage.

QUESTION 4.5 (Learning Outcome 5)

The question suggests that the aziridine ring opens in the following fashion:

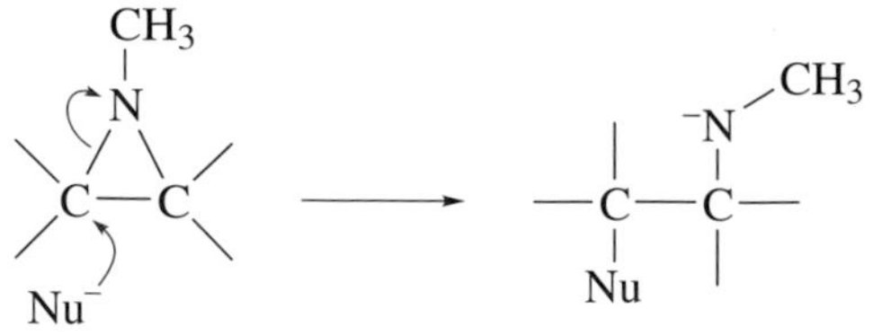

(Q.3)

A possible reaction for making pseudoephedrine would therefore be:

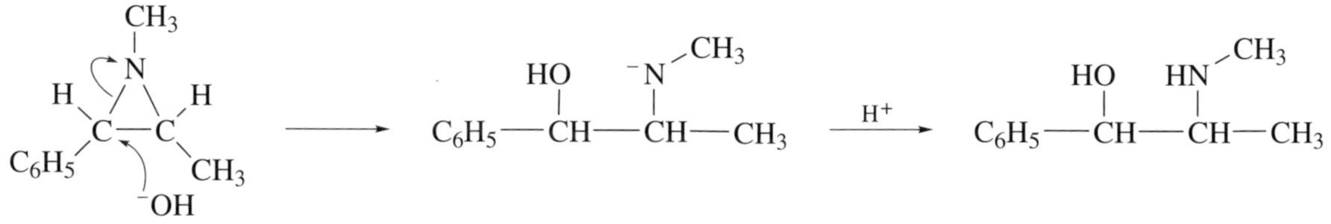

(Q.4)

However, because the question states that this will be an S_N2 reaction, steric hindrance will constrain the nucleophile ($^-$OH) to attack at the less sterically hindered site (that is, the carbon atom carrying a CH_3 group), giving the wrong product:

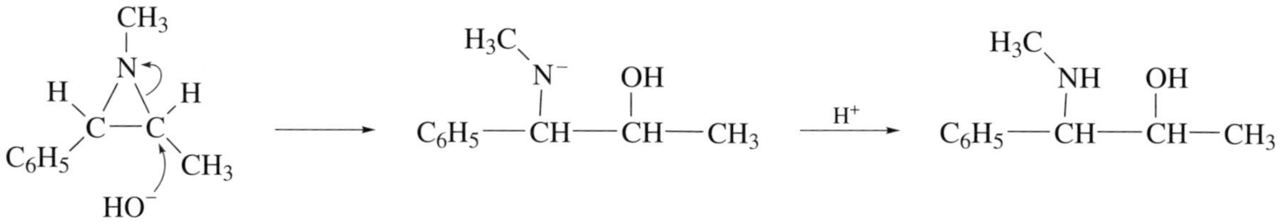

(Q.5)

So if this reaction were to proceed via an S_N2 mechanism, then this route would not be successful.

QUESTION 5.1 (Learning Outcome 7)

The molar mass of **5.21** is $147\,g\,mol^{-1}$, so a mass of $22.0\,g$ is equivalent to $0.15\,mol$ of the aziridine. If $0.15\,mol$ of **5.21** is used, we would expect a theoretical yield of $0.15\,mol$ of **5.22**. As **5.22** has a molar mass of $193\,g\,mol^{-1}$, the theoretical yield will be $0.15 \times 193 = 29.0\,g$ of product. $21.6\,g$ were obtained, so the percentage yield is:

$$\frac{21.6}{29.0} \times 100 = 74\%$$

QUESTION 5.2 (Learning Outcome 7)

The answer you obtained,

$$\frac{94}{100} \times \frac{93}{100} \times 100$$

should have been the same as the percentage yield calculated for the straight-through reaction discussed in the text — that is, 87%.

QUESTION 5.3 (*Learning Outcome 7*)

The overall yield is calculated by multiplying the yields (expressed as fractions) together, the final answer being expressed as a percentage:

$$\underset{\substack{\text{1st}\\\text{stage}}}{\frac{76}{100}} \times \underset{\substack{\text{2nd}\\\text{stage}}}{\frac{80}{100}} \times \underset{\substack{\text{3rd}\\\text{stage}}}{\frac{65}{100}} \times \underset{\substack{\text{last}\\\text{stage}}}{\frac{74}{100}} = 29\%$$

Although each yield looks reasonable, even with only four steps the *overall* yield is quite low.

QUESTION 6.1 (*Learning Outcomes 5 and 6*)

A possible synthetic route would be as shown in Scheme Q.1. The starting alkene could easily be prepared using an S_N2 reaction between the phenolic anion and 3-bromoprop-1-ene (allyl bromide). After that, the same procedure could be carried out, with propan-2-amine replacing methylamine in the last step. As the alkene does not exist in *E* and *Z* forms, only one set of enantiomers is possible. Again, resolution using a chiral acid would give the active enantiomer of propranolol.

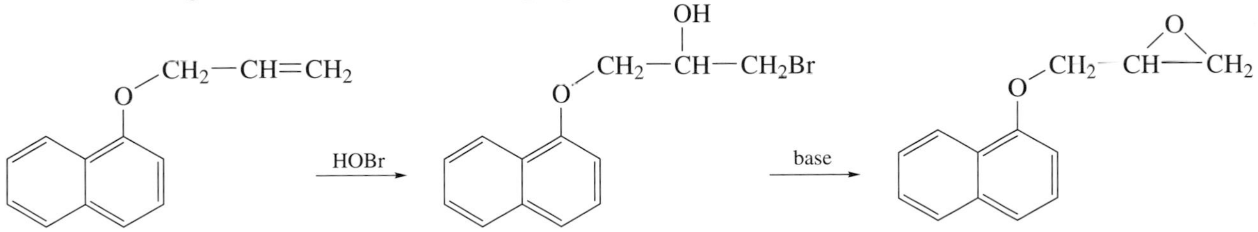

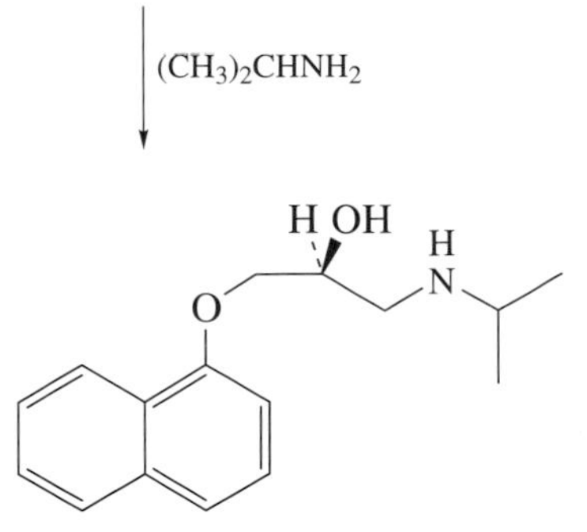

SCHEME Q.1

QUESTION 7.1 (*Learning Outcome 8*)

Addition looks to be the greenest of these three reaction types, as the resulting product has incorporated all (or most) of the atoms of the reagents involved. In substitution, some part of one of the reagents must be lost (the leaving group), as it is replaced by all, or part, of the other reagent. Elimination is the least green, since two components are lost from one of the reagents, and an additional reagent is usually needed to bring this about.

QUESTION 7.2 (*Learning Outcome 9*)

Your spreadsheet should show that the % atom economy for Route 1 is $100 \times 117/100 = 85\%$; an assessment table for it is shown below:

Step	Reactant Formula	FM/g	Utilized in methyl methacrylate Formula	FM/g	Not utilized in methyl methacrylate Formula	FM/g
1	C_3H_6O	58	C_3H_5O	57	H	1
1	CHN	27	C	12	HN	15
2	CH_4O	32	CH_3O	31	H	1
Total	$C_5H_{11}NO_2$	117	$C_5H_8O_2$	100	NH_3	17

As all the atoms of the reactants used in Route 2 are incorporated into the product, the % atom economy for it will be 100% — in other words, 15% more green than Route 1.

FURTHER READING

1   L. E. Smart and J. M. F. Gagan (eds), *The Third Dimension*, The Open University and the Royal Society of Chemistry (2002).

2   E. A. Moore (ed.), *Molecular Modelling and Bonding*, The Open University and the Royal Society of Chemistry (2002).

3   L. E. Smart (ed.), *Separation, Purification and Identification*, The Open University and the Royal Society of Chemistry (2002).

4   P. G. Taylor (ed.), *Mechanism and Synthesis*, The Open University and the Royal Society of Chemistry (2002).

ACKNOWLEDGEMENTS

Grateful acknowledgement is made to the following sources for permission to reproduce material in this book:

Figure 2.1: Jerry Mason/Science Photo Library; *Figure 2.2*: Damien Lovegrove/Science Photo Library; *Figure 2.3*: © The Nobel Foundation; *Figure 2.4*: Dr Jeremy Burgess/ Science Photo Library; *Figure 3.1*: courtesy of Pfizer.

Every effort has been made to trace all the copyright owners, but if any has been inadvertently overlooked, the publishers will be pleased to make the necessary arrangements at the first opportunity.

Case Study

Industrial Organic Chemistry

Alan Heaton

Liverpool John Moores University

1

INTRODUCTION

1.1 The chemical industry

In this Case Study we shall look at the large-scale production of the basic organic chemicals which constitute the starting materials for most of the output of the organic chemical industry.

Firstly, however, it is worth making some points about the industry as a whole, both organic and inorganic. Some figures for total production in 1995 are given in Table 1.1. The data were produced by the Royal Society of Chemistry, and though they are a little out of date, they give you a snapshot of how big the industry is world wide! It lies at the very centre of the manufacturing industry, with all other sectors utilizing its products to a greater or lesser degree. For example, modern motor cars (Figure 1.1) require large quantities of synthetic polymers — from the fabric for the seats to the dashboards, bumpers, steering wheels, and even the acrylic paint finish.

Figure 1.1
Examples of the plastics used in modern motor cars.

The contribution to other sectors may be much smaller, but is none the less significant — for example, providing chemicals for effluent treatment, and for the analysis of raw materials and finished products. In contrast to most other sectors, it is incredibly successful, having always had a very positive annual trade balance. Currently [2002], this stands at around £4 billion! It is thus an enormous wealth creator, with annual sales in excess of £40 billion, of which about 40% represents exports. It also provides employment directly for 250 000 people and indirectly for many more. The workforce is highly educated and skilled, with a significant proportion being graduates. Nevertheless, it is most important to appreciate that the chemical companies, like any other business, exist to make a profit, and their activities must be profitable if they are to stay in business.

Table 1.1 Production statistics of major chemicals (1995)

Common names	World production/Mt y^{-1}	Common uses
ABS copolymers (acrylonitrile–butadiene–styrene)	3.53	automotive industry; plastic bottles; electronic appliances
acetic acid (ethanoic acid)	6.44	production of vinyl acetate (ethenyl ethanoate) monomer; acetic anhydride (ethanoic anhydride); solvents
acetone (propanone)	3.4	production of methacrylates; bisphenol A; methyl iso-butyl ketone (4-methylpentan-2-one)
acrylonitrile (propenenitrile)	4	production of ABS; acrylic fibre; acrylonitrile–styrene copolymers
agrochemicals	$25.28 bn[1]	numerous uses
ammonia	114.22	fertilizers; explosives; fibres and plastics
benzene	32.4	production of ethylbenzene (for styrene monomer); intermediates for detergents and nylon
bromine	814 Mlbs[1]	ethylene dibromide (1,2-dibromoethane; anti-knock agent); bleaching; methyl bromide (fumigants)
butadiene	6.7	styrene–butadiene rubber; butadiene rubber; ABS resins
caprolactam	3.64	nylon 6; plastics; plasticizers
chlorine	40	production of chlorinated hydrocarbons; polyvinyl chloride (PVC); water purification
detergents (liquid)	$55–60 bn[1]	numerous uses
ethylene (ethene)	65.8	production of plastics (polythene or poly(ethene)); welding and cutting of metals
ethylene oxide (oxirane)	11.083	surfactants; fumigants; propellants; glycols
fertilizers	138	numerous uses
formaldehyde (methanal)	12	urea resins; phenolic resins; fertilizers; disinfectants; biocides
hydrogen peroxide	1.5	bleaching agent; pulp and paper; industry; rocket fuel
methanol	25	production of formaldehyde and acetic acid; solvents; chemical synthesis

Without the chemical industry our modern lifestyle would not be sustainable: there would be no computers, a much more limited range of colours and fabrics for clothes, furnishings, etc., and no non-stick cooking utensils. Indeed, our life expectancy would be much lower without the array of medicines and vaccines produced by the pharmaceutical sector.

Despite all these positive aspects, a high number of pollution incidents has given the industry a rather poor and negative public image (Figure 1.2), on a par with that of the nuclear industry! Of course, such problems attract considerable media attention, unlike the positive aspects that are seldom rated as newsworthy. In reality, this is slightly unfair, and the steps being taken to minimize or eliminate such problems in the future will be touched on towards the end of this Case Study.

Common names	World production/Mt y^{-1}	Common uses
unsaturated vegetable oils	6.48	soya oil; palm oil; sunflower and rapeseed oils
ortho-xylene (1,2-dimethylbenzene)	2.2	production of phthalic anhydride; dyes; insecticides
para-xylene (1,4-dimethylbenzene)	9.7	production of terephthalic acid (polyester fibres and resin); insecticides
phosphoric acid	100	numerous uses
polycarbonates	2.2 billion lbs[1]	moulded products; extruded film; non-breakable windows
polyethylene (poly(ethene))	37	low-density polyethylene (LDPE); high-density polyethylene (HDPE)
polyethylene terephthalate (PET)	2.6[2]	soft drink bottles; packaging films; recording tapes
polypropylene (poly(propene))	20.7	packaging films; kitchenware; automobile parts; fibres
polystyrene (poly(phenylethene))	12.045	packaging; thermal insulation; furniture construction
polyvinyl chloride (PVC, poly(chloroethene))	18.633	piping and conduits; plumbing and construction; electrical insulation
propylene oxide (methyl oxirane)	4.048	propylene glycols; surfactants and detergents
purified terephthalic acid	11.5	polyester fibres; PET bottle resins; films
sodium hydroxide	39	chemicals manufacture; rayon and cellophane; neutralizing agent; detergents; soaps and textiles
styrene (phenylethene)	15.729	polystyrene; various resins (including ABS)
sulfuric acid	135.67	chemicals manufacture; fertilizers; industrial explosives
titanium dioxide	3.7	white pigment in paints; paper industry; cosmetics
toluene (methylbenzene)	17.6	production of benzene and xylene; aviation fuel; adhesive solvents
vinyl chloride monomer (chloroethene)	19.34	polyvinyl chloride (poly(chloroethene), PVC)

1 Only available figure.   2 Excluding fibres and films.

Figure 1.2   The popular, negative view of the chemical industry.

It is worth pointing out that the general public does not normally come across chemicals as such, but rather meets them several steps further down the manufacturing process chain, when they have been utilized in consumer products; examples are paints, drugs, pesticides, and synthetic fibres such as polyesters.

Let us take a specific example to illustrate this. Poly(vinyl acetate), or, using its systematic name, poly(ethenyl ethanoate), is an important component of emulsion paints. Its function is to bind the pigment (titanium dioxide) such that a homogeneous paint film is produced when the water base evaporates. Production of the polymer starts from crude oil, and after this has been fractionally distilled, a suitable fraction such as *naphtha*, is then 'cracked' (Section 2.4) to give ethylene (ethene).*

* Although we have used systematic nomenclature for compound names in the earlier parts of this Book, the older trivial names are the norm in the chemical industry, which is why the latter names are used in this Case Study.

The ethylene (ethene) is reacted with acetic (ethanoic) acid and oxygen over a supported palladium catalyst to produce vinyl acetate.

$$\underset{H}{\overset{H}{\diagdown}}C=C\underset{H}{\overset{H}{\diagup}} \quad + \quad H_3C-C\underset{OH}{\overset{O}{\diagup}} \quad \xrightarrow[-H_2O]{Pd/O_2} \quad H_3C-C\underset{O-CH=CH_2}{\overset{O}{\diagup}} \tag{1.1}$$

This is then polymerized. Finally, the poly(vinyl acetate) is mixed with the other ingredients to produce the emulsion paint.

1.2 Large- and small-scale production

What do 'large' and 'small' scale mean in the context of organic chemicals manufacture? A good example of a small-scale process is the synthesis of a drug. Products such as this are typically produced on the 10–100 tonnes *per annum* scale (1 t = 1 000 kg). In contrast, a single plant producing ethylene (ethene) and other products by cracking could process 600 000 tonnes *per annum* of feedstock. This is an example from the petrochemicals sector of the chemical industry, and it is here that almost all the large-scale organic chemical processes are found.

1.3 Sub-division of the organic chemicals industry

The chemical industry may be subdivided into two sectors: (a) petrochemicals and (b) speciality and fine chemicals. For each of these, we shall look at the particular characteristics, the sort of chemistry involved, and the constraints on the location of the manufacturing plants. In addition, for petrochemicals we shall consider the reasons why oil and natural gas are the source of more than 90% of all organic chemicals. We shall also address the theoretical considerations of producing chemicals from these sources, plus the environmental consequences of having a thriving chemical industry.

PETROCHEMICALS

2

As the name implies, petrochemicals are derived from petroleum or crude oil (and natural gas). Like its sister industry, oil refining, its operations are carried out on an enormous scale, and its plants are among the giants of the chemical industry. Single plants, such as that shown in Figure 2.1, can process 600 000 tonnes *per annum* of feedstock. Assuming round-the-clock working, that means greater than 10 000 tonnes per week or 1 500 tonnes per day! The majority of the plants in this sector would have a capacity of at least 100 000 tonnes per annum.

2.1 Theoretical considerations

Crude oil consists mainly of alkanes and cycloalkanes, plus much smaller amounts of other compounds such as alkenes and aromatics. Natural gas is mainly methane, plus a little ethane.

Figure 2.1
Large petrochemicals plant at Grangemouth.

● Would alkanes and cycloalkanes be a logical starting point for making a variety of other organic compounds?

● No, alkanes are just about the least reactive of organic compounds, whereas what is required in synthesis are compounds with reactive functional groups.

Nevertheless, you should remember that nowadays more than 90% of all organic chemicals are manufactured from crude oil and natural gas. So the alkanes have to be transformed into more reactive compounds.

● Which other compounds are fairly similar in structure to alkanes but much more reactive?

● Alkenes also contain only carbon and hydrogen atoms, but they have a reactive carbon–carbon double bond.

In a process known as **cracking**, suitable crude oil fractions may be converted into alkenes. A related process called **reforming** enables the alkanes to be turned into aromatic compounds. These conversions of the more-stable saturated alkanes into the unsaturated, more-reactive alkenes and aromatics come at a price. They require considerable amounts of energy. However, the starting materials (reactants or feedstocks) are relatively cheap.

Nevertheless, at first sight it may seem surprising that crude oil/natural gas has become *the* source of organic chemicals. In fact, this strategy is relatively recent: in 1950, 60% of organic chemicals manufactured in the UK were obtained from coal, 31% from carbohydrates and only 9% from oil; nowadays, the dominance of oil as a feedstock is almost complete, as you can see from Figure 2.2.

To obtain organic chemicals from coal, it first had to be *carbonized* — that is, heated strongly in the absence of air. This gave products such as gas, crude benzole (benzene) and coal tar (Part 2 Section 1). Further processing of the coal tar afforded aromatic chemicals such as benzene, toluene and phenol. The gas was used to produce town gas (a mixture of principally hydrogen and carbon monoxide), which was used as a domestic and industrial fuel. Coke was by far the major product of the carbonization, and in the 1950s the demand for it was high, both as a smokeless fuel and in steel-making. However, in recent times demand for coke has fallen substantially, and it is no longer economic to carbonize the coal purely for the chemicals. Demand for town gas has also ceased since it was replaced by natural gas.

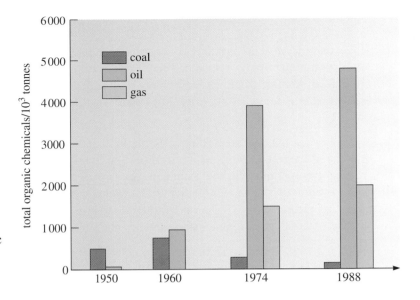

Much research is continuing into alternative ways of obtaining chemicals from coal. The methods can be split into (a) gasification and (b) liquefaction, both of which produce products that are similar to either natural gas or crude oil fractions.

Figure 2.2
Sources of organic chemicals in the UK from 1950 to 1988.

Nevertheless, the feedstock for the organic chemical industry has changed from coal to crude oil, mainly, because oil became abundant, cheap and readily available in large quantities (Figure 2.3), but also because, being a liquid, it was easier to 'handle' and transport. In contrast, coal, having to be mined, was harder to obtain, and as a solid was more difficult to handle; for example, it couldn't be easily pumped around industrial plants.

Figure 2.3
An oil-drilling platform.

Petrochemical companies operate on a gigantic scale, and are able to take full advantage of the 'economy of scale effect'. Put simply, this states that the more of anything that is produced, the cheaper each unit (or tonne) costs to produce. The classic example is the motor vehicle industry, where factories have been expanded to produce more and more cars per hour. For chemicals, the relationship is shown in Figure 2.4.

Actual figures for the cost are deliberately not included in Figure 2.4; it is the shape of the graph which is important. It shows a reduction in cost per tonne as the plant capacity is increased. However, eventually there is no further reduction to be obtained by increasing the capacity. This is one of the reasons why no plant bigger than 600 000 tonnes per annum capacity has ever been built.

In concluding this Section, it is important to note that it is possible to produce the majority of industrial organic chemicals from *any* of the sources already mentioned, namely oil, natural gas or coal; the current dominant position of oil is largely a matter of economics.

2.2 Characteristics of the petrochemical sector

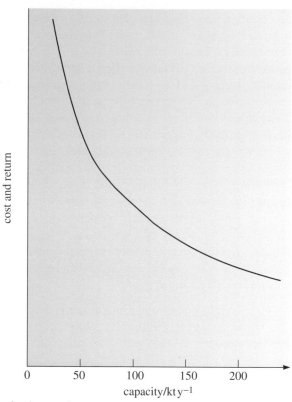

Figure 2.4
The relationship between unit cost and capacity of production.

Petrochemicals have a low value per unit mass — typically, several hundred pounds sterling per tonne; we have already noted that they are produced on a vast scale. They are therefore made in plants dedicated to a single product, operating on a *continuous* basis and computer controlled. Thus, reactants are continuously fed into one part of the reactor while products are being continuously removed from another part of it. Each plant represents an enormous capital investment in technology. As an illustration, Shell spent over £100 million on their SHOP (Shell Higher Olefins [alkenes] Plant) in the 1970s. The plants are often run deliberately at low conversion rates, with a lot of unreacted materials being recycled. This is done to minimize byproduct formation, and to increase the yield of the desired product. It also reduces the separation costs for isolating the desired products.

The importance of obtaining a high yield in the petrochemical sector cannot be overemphasized. This is mainly because the (relatively) low price commanded by the products means that there is a small profit margin. Hence, a very small increase in yield can have a dramatic effect on revenues if the scale of operation is very large: a one per cent improvement could easily generate an extra £1 million per annum! This is why companies in the petrochemical sector are constantly striving for very high reaction yields; they are often able to achieve yields in excess of 90%.

2.3 Energy considerations

The chemical industry as a whole, and petrochemicals in particular, are major energy users. Energy is required for heating reactors, for carrying out reactions under high pressure, and also for separating the desired product(s) from byproducts. In petrochemicals, the last of these operations usually means resorting to one or more distillations (Figure 2.5).

Figure 2.5 Distillation tower used in the refining of crude oil.

Note also that the petrochemicals industry competes for its raw materials with the energy producers, since most energy is derived from crude oil fractions and natural gas.* Any increases in the price of energy therefore hit the chemical industry twice. In the past two decades, great effort has been devoted to improving energy efficiency and conservation, and this has become an important aspect in the design of new plants (see also the account of 'green chemistry' in Part 3 Section 7.1). This change in focus was triggered by the more than quadrupling of oil prices in 1970 by OPEC (Organization of Petroleum Exporting Countries).

2.4 Chemical considerations

The starting point for the petrochemicals industry is crude oil, which, after removal of any sulfur, is subjected to distillation. As crude oil is a complex mixture consisting largely of alkanes and cycloalkanes (Table 2.1), this involves separation into several less complex mixtures. The compounds in crude oil are chemically similar, so they are separated according to their boiling temperature. The boiling temperatures of alkanes increase with molecular mass, which means that separation essentially takes place according to the number of carbon atoms in the molecule, as shown in Table 2.2. The lower-boiling fractions are the most useful.

---

* Only about 8% of crude oil is used for chemicals manufacture.

Table 2.1 Composition by mass of crude oils

Hydrocarbon class	All crude oils (approximate percentage by mass)	North Sea oil (approximate percentage by mass)
straight chain	20–65	40
iso (2-methyl)	5–20	5
cycloalkane	20–50	35
aromatic	10–40	20

Table 2.2 The boiling fractions from the distillation of crude oil

Fraction	Boiling range/°C	No. of carbon atoms in molecules	% of crude oil (v/v*)
gases	< 20	1–4	1–2
light naphtha (gasoline)	20–70	5–6	20–40
naphtha	70–170	6–10	
kerosene	170–250	10–14	10–15
gas oil (diesel)	250–340	14–19	15–20
lubricating oil	340–500	19–35	15–20
waxes	340–500	19–35	2–3
bitumen	>500	>35	25–30

* 'v/v' means percentage by volume.

The industrial plant needed for the fractionation of crude oil is very different in appearance from that used in a laboratory for similar separations (Figure 2.6a)*. The distinctions are largely a matter of scale, however. A schematic representation of a fractionation tower is shown in Figure 2.6b. The oil is heated very strongly by gas burners, using gas obtained as a byproduct of the fractionation. The hot, vaporized oil then passes up the tower, where it encounters a series of trays. The higher-boiling fractions collect in the lower trays, and are taken off in pipes; the lower-boiling fractions collect in the higher trays before being taken off. Gases such as methane, ethane and propane are led off from the top of the tower to be returned to the burners. The boiling range of the fractions can be controlled very accurately by adjusting the amount of heating in the first stage.

The next step is to turn the alkanes in these fractions into alkenes by *cracking*, and into aromatics by *reforming*. A brief, and *very simplified*, outline of these processes is presented here.

* Techniques for separation are discussed in *Separation, Purification and Identification*[1].

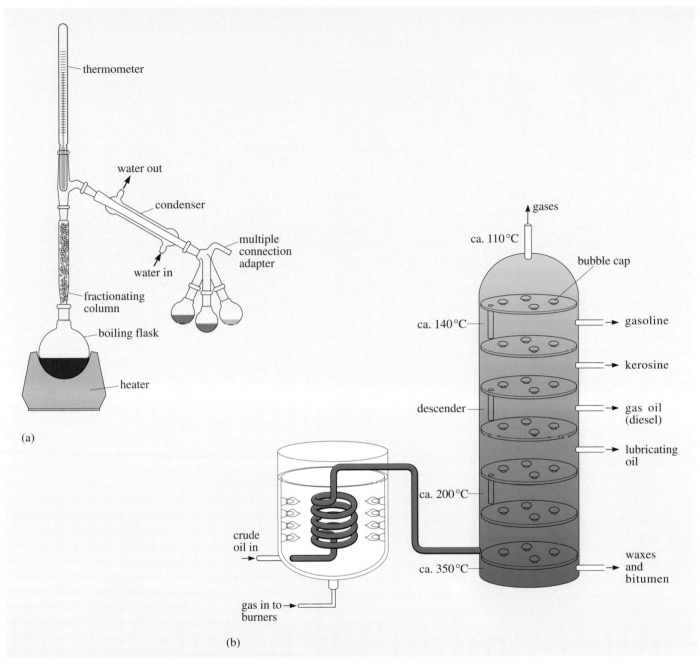

Figure 2.6 (a) Laboratory distillation using a fractionating column. (b) Schematic representation of a fractionation tower for oil distillation. A typical tower may be 60 metres tall. Vapour from one tray passes to the one above, where it condenses in the bubble caps. Liquid in one tray falls to the tray below through a descender pipe.

2.4.1 Cracking

The cracking of a typical alkane, hexane, can be represented as follows:

$$2C_6H_{14} \xrightarrow[\text{catalyst}]{500\ °C;\ 2\ \text{bar}} CH_4 + 3C_2H_4 + C_2H_6 + C_3H_6 \qquad (2.1)$$

● Identify which of these hydrocarbons are alkanes, and which are alkenes.

○ CH_4 (methane) and C_2H_6 (ethane) are alkanes, and C_2H_4 (ethylene (ethene)) and C_3H_6 (propylene (propene)) are alkenes.

The process is carried out in a catalytic cracker, where the catalyst is a zeolite*. In reality, the feedstock is not a single compound such as C_6H_{14} but one of the crude oil fractions. The fraction that is used as a raw material varies from one refinery to another, and is often the fraction that is readily available and surplus to fuel needs. Often the starting material is a lubricating oil, the wax fraction, or the naphtha fraction.

Table 2.3 shows the distribution of products obtained from different feedstocks. As we would expect, it shows that the longer the carbon chains in the feedstock, the greater the proportion of longer chains in the products. For example, ethane gives 75–80% ethylene (ethene), whereas naphthas give only 25–37% of ethylene. Unreacted alkanes are recycled. There is much less demand for the high-boiling fractions, so cracking is used to turn them into naphtha-like materials.

Table 2.3 The percentage proportion of products obtained by the catalytic cracking of different fractions of crude oil

| | | | Feedstock | | |
Product	Ethane C_2	Propane C_3	Butane isomers C_4	Naphthas C_5–C_{13}	Gas oil C_{14}–C_{19}
hydrogen	5–6	1–2	1	1	0.5
methane	10–12	20–25	15–25	13–18	10–12
ethylene (ethene)	75–80	40–45	20–30	25–37	22–26
propylene (propene)	2–3	15–20	15–25	12–16	14–16
butadiene	—	1–2	0–2	3–5	3–5
other C_4s	1	3–4	15–22	3–5	3–6
higher hydrocarbons (b.t. 20–220 °C)	1	5–10	5–14	18–28	17–22
fuel oil	—	—	—	4–8	18–22

* The structures and applications of zeolites are discussed in the Case Study in *Chemical Kinetics and Mechanism*[2].

2.4.2 Reforming

Reforming can be represented in an oversimplified form as:

$$C_6H_{14} \xrightarrow[\text{catalyst}]{500\,°C,\ 25\,bar} \bigcirc + 4H_2 \tag{2.2}$$

A typical catalyst comprises platinum or a platinum (99–99.5%)–rhenium (0.5–1%) alloy on an acidic alumina support. The reactions occurring include cyclization and dehydrogenation.

Again, in reality, the feedstock is a mixture of compounds; Table 2.4 shows a typical feedstock and product composition. The distribution of aromatic compounds in this reformed product is given in Table 2.5. You will observe that 'sod's law' operates, in that benzene is the product that is in most demand, yet it is only 4% of the product! This is no problem though for the research chemists, who have developed processes to dealkylate toluene (methylbenzene) and the xylenes (dimethylbenzenes), and convert them into benzene.

Table 2.4 The composition of a typical feedstock for catalytic reforming and the distribution of products formed

	Feedstock	Product (ca 75% yield by volume)
paraffins (alkanes)	60	32
naphthenes (cycloalkanes)	25	2
aromatics	15	66

Table 2.5
The composition of the aromatics fraction in the product of Table 2.4

Product	Percentage
benzene	4
toluene	18
xylenes and ethylbenzene	23
higher aromatics	21

2.4.3 Building blocks

These two processes — cracking and reforming — are used to produce the *six* key intermediates, or 'building blocks' (counting the three xylenes as *one* building block), from which the majority of industrially produced organic chemicals are made. They are:

$$H_2C{=}CH_2 \qquad CH_3CH{=}CH_2 \qquad CH_2{=}CH{-}CH{=}CH_2$$

ethylene   propylene   buta-1,3-diene

benzene   toluene   1,2-dimethylbenzene (*ortho*-xylene) ⌨   1,3-dimethylbenzene (*meta*-xylene) ⌨   1,4-dimethylbenzene (*para*-xylene) ⌨

the xylenes

In simple terms, many petrochemical processes can be summed up as: passing reactant(s) over a catalyst in a hot tube (400–500 °C), with the products emerging at the other end! Some reactions also involve high pressures (up to 200 bar). The hard part is finding an appropriate catalyst and suitable conditions! The reactions are therefore energy demanding, emphasizing the earlier comments on the importance of energy conservation and efficiency. However, one advantage is that at the high temperatures the reactions are very fast; sometimes they are complete in less than 1 second! Clearly, this aids the high throughput of materials, which is needed to yield the enormous quantities required of these chemicals.

Figures 2.7, 2.8, 2.10, 2.11, 2.13 and 2.15 indicate *some* of the products which are manufactured from each of the six key building blocks. Many of these products are made in plants having capacities of 100 000 tonnes per annum. These charts are provided to impress on you the amazing versatility of the feedstocks, and the wide range of products that each one yields.

2.4.4 Ethylene (ethene)

For many years ethylene (ethene) has been the most important organic compound, as shown by Table 1.1. Figure 2.7 clearly demonstrates why ethylene (ethene) is needed in such large quantities, since several polymers and polymer intermediates are made from it. Production of each of these exceeds 1 million tonnes per annum. Examples are polyethylene, PVC (poly(vinyl chloride)), PVA (poly(vinyl acetate)) and polystyrene. In fact, in tonnage terms, over 60% of all industrial organic chemicals are used in the manufacture of synthetic polymers.

2.4.5 Propylene (propene)

As explained previously, propylene (propene) was originally just a waste byproduct in the preparation of ethylene (ethene) by cracking. Figure 2.8 (overleaf) bears testimony to the ingenuity of research chemists in developing all these uses and processes to utilize propylene.

Acrylonitrile and acrylic acid are turned into polymers — for example, Acrilan™ and Enkalon™ (Figure 2.9) — and acrylic paints. Propylene dimer has been used to make the transparent plastic poly(4-methylpent-1-ene). The trimer is used in the manufacture of synthetic lubricants. In contrast, the tetramer is used to manufacture synthetic detergents such as sodium dodecylbenzenesulfonate, or ABS (alkylated benzenesulfonate) as it is known. This formed the basis of the heavy-duty washing powders developed in the 1950s and 1960s. Its synthesis (shown in Scheme 2.1) demonstrates the application of the standard aromatic chemistry discussed in Part 2 .

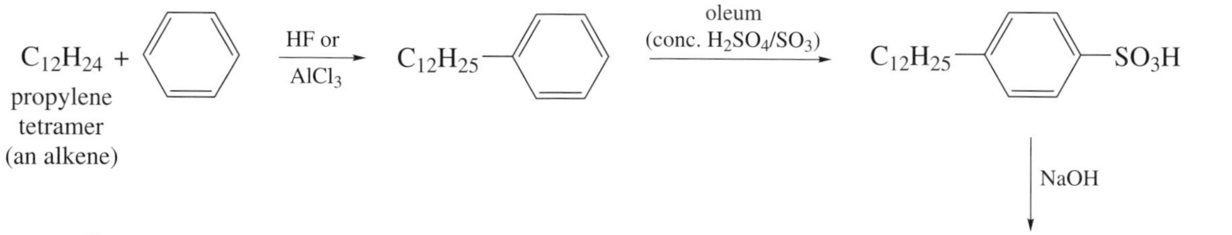

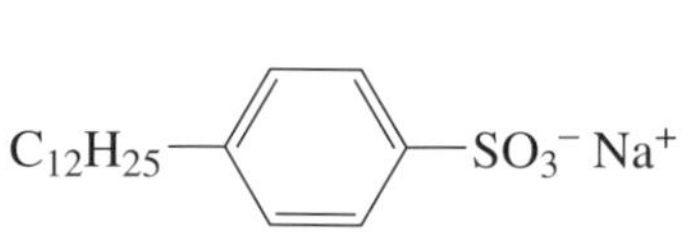

$$C_{12}H_{24} + \text{⬡} \xrightarrow[\text{AlCl}_3]{\text{HF or}} C_{12}H_{25}\text{—⬡} \xrightarrow[\text{(conc. H}_2\text{SO}_4\text{/SO}_3)]{\text{oleum}}$$

propylene
tetramer
(an alkene)

2.4.6 Buta-1,3-diene

As shown in Figure 2.10 (overleaf), most buta-1,3-diene is polymerized, either on its own or with other alkenes to give co-polymers like styrene–butadiene rubber, SBR. The major use of polybutadiene and SBR rubbers is for car tyres.

Poly(acrylonitrile–butadiene) is an example of a nitrile rubber. It has outstanding resistance to oils and abrasions — hence its use for oil seals, flexible fuel tanks in aircraft, oil-resistant hoses and for ink rollers for printing presses. Polymerization of

Figure 2.7
Intermediates and products from ethylene (ethene); note the widespread use of catalysts in these reactions. (In this and the similar Figures that follow, the intermediates are shown on red (first) and blue backgrounds), and the products are shown on green backgrounds.) 💻

Figure 2.9
Bottles made from acrylonitrile.

$$C_{12}H_{25}\text{—⬡—}SO_3H$$

$$\downarrow \text{NaOH}$$

$$C_{12}H_{25}\text{—⬡—}SO_3^- Na^+$$

SCHEME 2.1

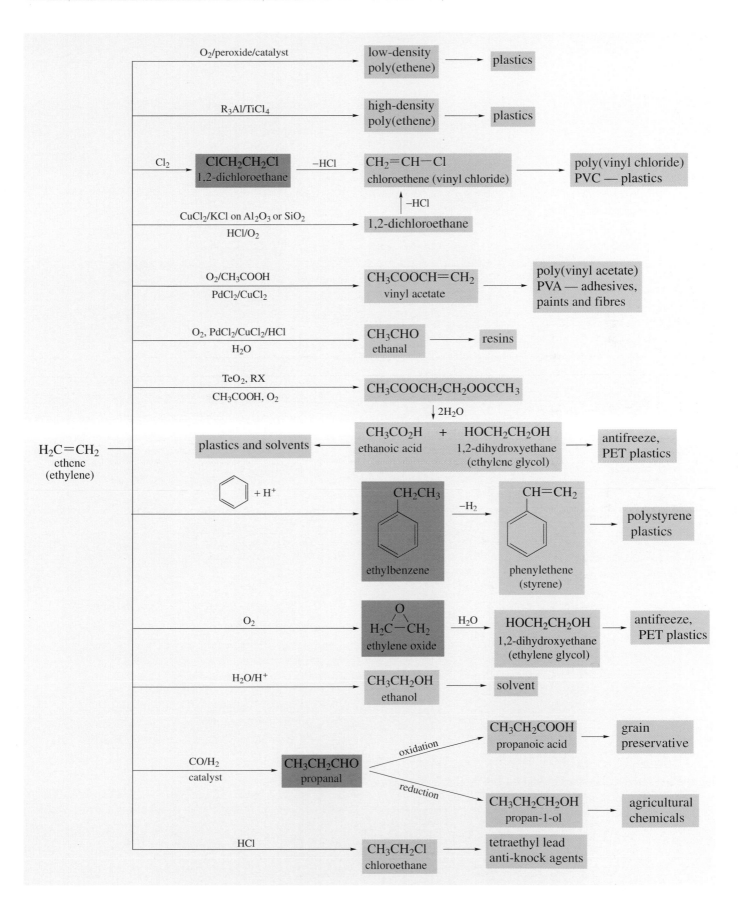

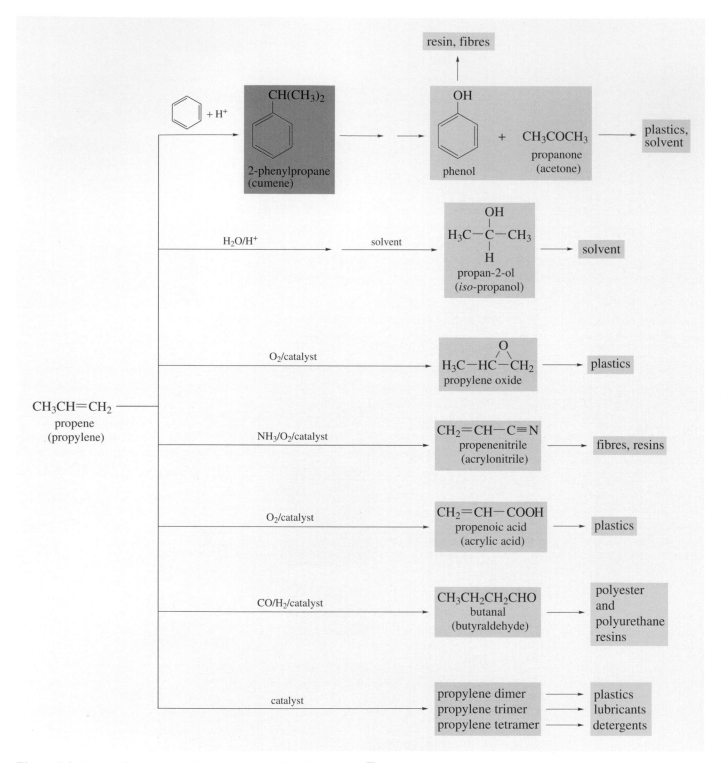

Figure 2.8 Intermediates and products from propylene (propene).

chloroprene produces a neoprene rubber. This has excellent resistance to solvents and heat.

It therefore finds uses in hoses for chemicals and in car engines, and also in the manufacture of chemically resistant laboratory gloves.

Reaction of buta-1,3-diene with hydrogen cyanide, HCN, gives adiponitrile (hexane-1,6-dinitrile). This is reduced to hexamethylenediamine (hexane-1,6-diamine, $H_2N(CH_2)_6NH_2$), which is one of the reactants used to make nylon-6,6 (**2.1**).

2.1

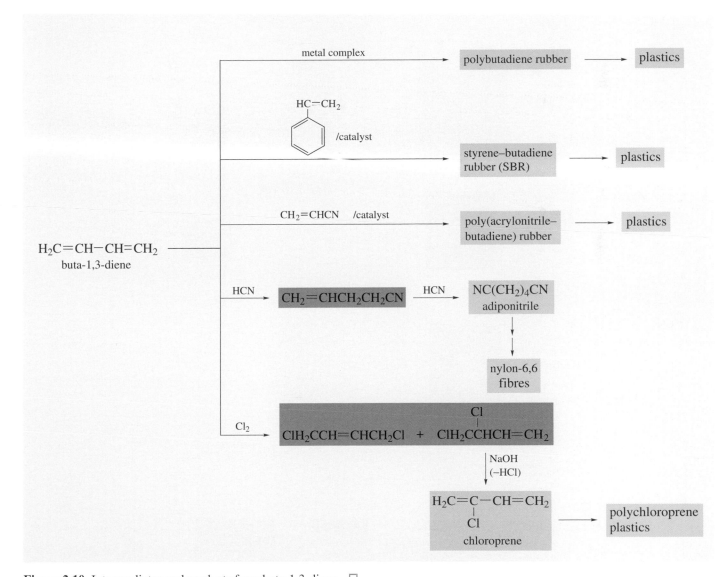

Figure 2.10 Intermediates and products from buta-1,3-diene. 💻

2.4.7 Benzene

The two major uses of benzene (Figure 2.11) involve reaction with two of the other key intermediates. These are, firstly, a reaction with propylene to give cumene (*iso*-propylbenzene; it is also called 1-methylethylbenzene or 2-phenylpropane, Figure 2.8), which is converted into acetone (propanone) and phenol. Secondly, there is a

Figure 2.11
Intermediates and products from benzene. 💻

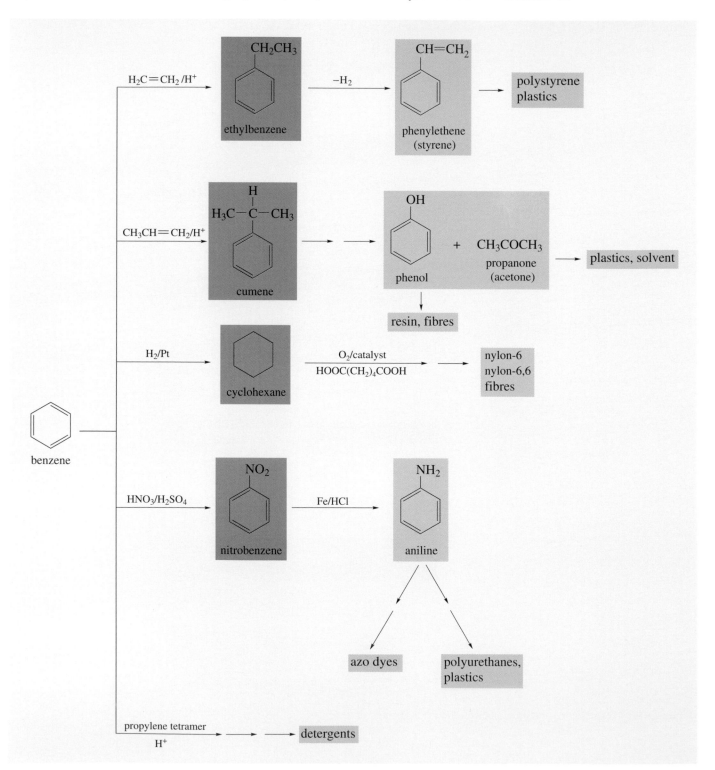

Figure 2.12
Polystyrene packaging.

similar reaction using ethylene (ethene) in place of the propylene, which affords ethylbenzene; this is then dehydrogenated to form styrene (phenylethene; see Figure 2.7). The major use of both phenol and styrene is to make polymers (Figure 2.12).

Complete reduction of benzene affords cyclohexane. Vigorous oxidation of cyclohexane gives adipic acid (hexane-1,6-dioic acid, $HOOC(CH_2)_4COOH$, which is the other component used in the manufacture of nylon-6,6 (**2.1**).

2.4.8 Toluene (methylbenzene)

In addition to its use as a solvent, toluene can be dealkylated to give benzene, as shown in Figure 2.13 (overleaf). It is a precursor for phenol, via oxidation to benzoic acid. However, this is a minor route compared to that discussed earlier from benzene via cumene. The vast majority of phenol is now made via cumene, and all recent plants use this route. It has the advantage that the other product, acetone, is also required in huge quantities. Nevertheless, it is only one of five routes to phenol that have been used commercially, which is testimony to the importance of phenol.

The major use of toluene is in the synthesis of toluene diisocyanate, TDI (see Figure 2.13), which is required for polyurethane production (Figure 2.14).

Figure 2.14
A desk coated with a polyurethane varnish.

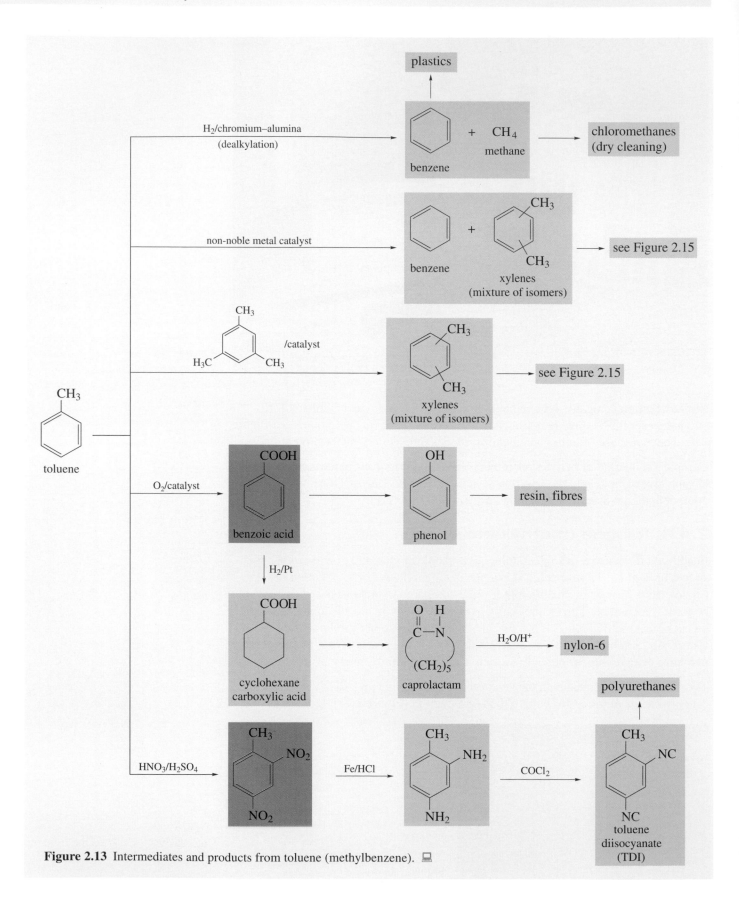

Figure 2.13 Intermediates and products from toluene (methylbenzene). 💻

2.4.9 Xylenes (dimethylbenzenes)

The xylenes are dimethylbenzenes. Three isomers are possible, known as *ortho-*, *meta-* and *para*-xylene. Catalytic reforming affords a mixture of all three, which can be separated, although it is not easy. This is because their boiling temperatures are very similar: *ortho* 144 °C; *meta* 139 °C; *para* 138 °C. Using very tall, highly efficient distillation columns enables the *ortho-* to be separated from the mixture of *meta-* and *para*-isomers. Fractional crystallization of the latter mixture at –60 °C causes the *para*-isomer to crystallize out, leaving the *meta*-isomer in the residual liquor (*para*-xylene is also obtained using shape-selective catalysis over zeolites*).

As Figure 2.15 shows, the main fate of the xylenes is oxidation to diacids and phthalic anhydride. These are then turned into polyesters such as Terylene™ and other polymers.

This Section has shown just *some* of the chemicals which can be manufactured from the six key building blocks. Note that quite a number of these products are themselves made on an enormous scale — that is, in excess of 1 million tonnes per annum. Most of the production ends up being turned into synthetic polymers, and this has been the driving force for the enormous increase in the output of organic chemicals since the 1950s.

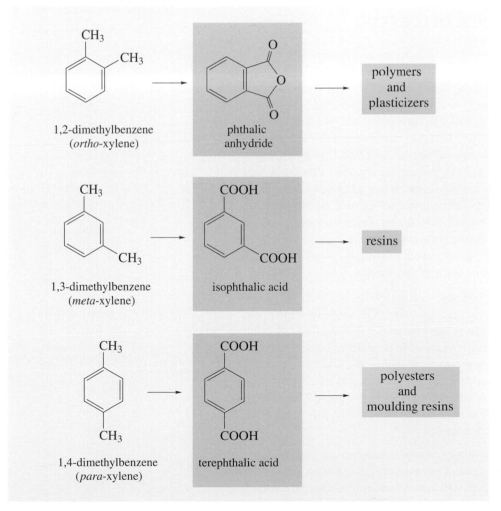

Figure 2.15
Intermediates and products from the xylenes.

* Zeolites are the subject of the Case Study in *Chemical Kinetics and Mechanism*[2].

BOX 2.1 A new petrochemical star

In the last few years, production of one particular organic chemical has increased in leaps and bounds, such that it is now challenging ethylene's premier position, and may soon overtake it. It is MTBE — methyl tertiary butyl ether (2-methoxy-2-methylpropane). The phasing out of leaded petrol has meant the loss of tetraethyl lead as an anti-knock agent. Thus, another compound had to be found that would improve the octane rating (based on the performance with respect to 2,2,4-trimethylpentane defined as 100) of petrol so that it would burn smoothly in car engines. MTBE is an excellent octane enhancer, and as use of unleaded petrol increases, more MTBE will be required. MTBE is relatively cheap to manufacture because it is made from methanol and isobutylene (2-methylpropene):

$$CH_3OH \ + \ \begin{matrix} H_3C \\ \diagdown \\ C=CH_2 \\ \diagup \\ H_3C \end{matrix} \longrightarrow \begin{matrix} H_3C \diagdown \diagup CH_3 \\ C \\ CH_3-O \diagup \diagdown CH_3 \end{matrix} \qquad (2.3)$$

2.5 Environmental concerns

The general public generally sees the chemical industry as a significant source of environmental pollution. Although there are specific examples of this, the case has been overstated. There is no doubt that the industry has for some years been making major efforts to improve the situation. Two examples can be cited to illustrate this. Firstly, considerable efforts are now devoted to 'designing out' waste when new plants are at the planning stage. Clearly, it is better, and more efficient, not to create the waste in the first place. If this is not possible, the BATNEEC (best available techniques not entailing excessive costs) system can be applied in the treatment of the waste. Bear in mind that while complying with the standards for effluent discharge, the company is still trying to achieve maximum profitability. In this respect, it is worth noting that the chemical industry has been very successful in utilizing its byproducts, rather than allowing them to become waste. Propylene is an excellent example of this, as Figure 2.8 showed; originally, propylene was just burned as a waste byproduct of ethylene (ethene) production.

Secondly, 'green chemistry' (Part 3 Section 7) is rapidly catching on. This uses alternative chemistry to replace existing reactions which produce unwanted and sometimes toxic waste. A good example is the Friedel–Crafts reactions you met in Part 2. These traditionally use aluminium trichloride or another Lewis acid catalyst. The work-up produces highly acidic solutions containing aluminium salts. These can now be avoided by using 'Envirocats™', which are supported reagent catalysts (such as $AlCl_3$ on clay), and are non-toxic, non-corrosive and inert. They can also be used in ways in which no aqueous effluent is produced. Clearly, processes utilizing them will be more environmentally friendly.

Waste that has to be disposed of represents a loss to the chemicals manufacturer in two ways. Firstly, it represents loss of raw material that has not been converted into the desired product. Secondly, the waste will almost certainly require treatment before it can be discharged or disposed of. Both of these factors will increase the cost of making the desired product.

Just like any other industry, the chemical industry is trying to be as profitable as possible, so as to give maximum return to its shareholders and other stakeholders. Thus, it is unlikely to embrace any expensive environmentally friendly activity without some persuasion. Recent environmental legislation (and environmental lobbying) has ensured that the chemical industry is now looking at ways in which it can reduce emissions. If this can be achieved with savings on consumable and running costs, it could be very attractive for the industry. One way of achieving this was discussed in Part 3 Section 7.

2.6 Location of plants

All major petrochemical plants are located on river estuaries. Examples are Grangemouth, Merseyside, Southampton and Teesside. This is because very large amounts of feedstocks and chemical products are involved. Transportation of these relatively low-value materials is expensive, and should be minimized. It is therefore logical to site them next to their raw material suppliers — the oil refineries. These tend to be located on river estuaries because crude oil is traditionally transported by sea (Figure 2.16). In addition, other plants that utilize the large output of their primary products will form part of the petrochemicals complex. An example would be a plant that polymerized ethylene (ethene) to poly(ethylene).

Figure 2.16
An oil tanker at sea.

An additional factor is that the plants will certainly require large quantities of cooling water — for example, to pass through the coils of a condenser in a distillation unit or in some form of heat exchanger. Such water is readily available from the river.

The company will also require an outlet for discharging the treated effluent which the plants have produced. A river, particularly near to where it joins the sea, is ideal. The tidal estuary and sea will help to dilute and disperse the material.

A further advantage is that the river and sea allow bulk transportation of raw materials and finished products. We saw at the beginning of this Case Study that around 40% of chemicals production in the UK are exported (p. 140) — hence the importance of such locations.

Note that transportation costs can be a significant proportion of total costs; this contrasts with the situation that obtains with speciality and fine chemicals, as we shall see in the next Section.

SPECIALITY AND FINE CHEMICALS

3

3.1 Some typical fine chemicals

As the name implies, fine chemicals are produced for a particular or 'special' application. They encompass agrochemicals (pesticides, or crop protection agents, as they are now described — a much better PR term!), pharmaceuticals, dyestuffs and speciality polymers. A few specific products from these areas are shown in Figure 3.1. Clearly, they are much more varied and complicated than the primary products of the petrochemical industry. We shall look at each in turn.

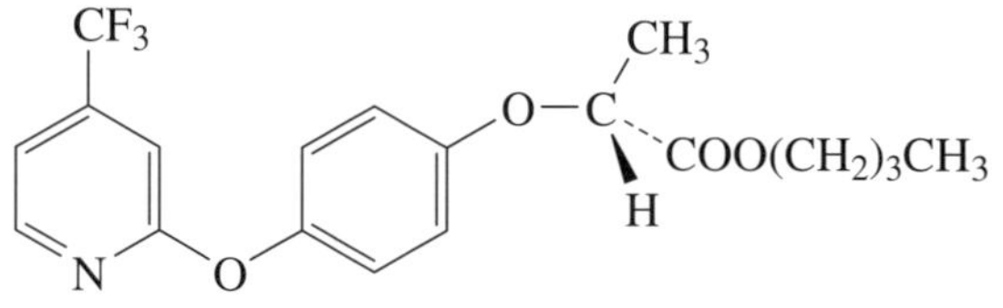

fusilade™

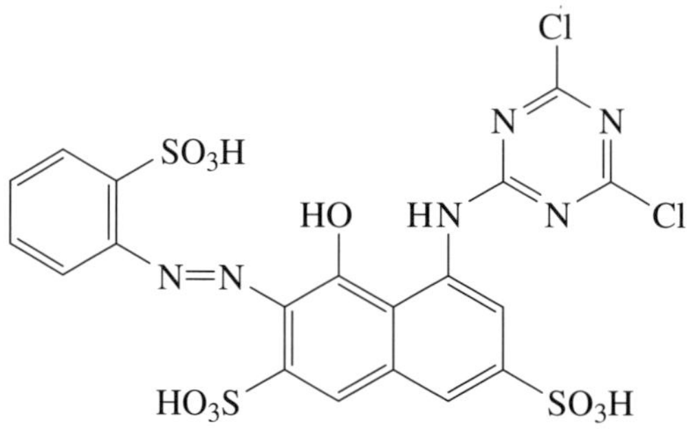

procion dye

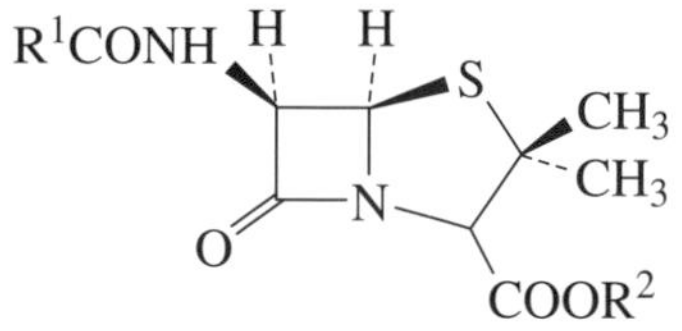

penicillins

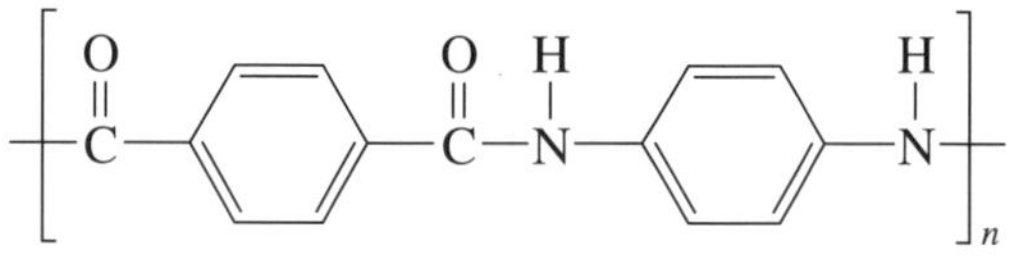

Kevlar

Figure 3.1 Some typical fine chemicals.

3.1.1 Fusilade™

Fusilade™ is a modern selective herbicide, used to control grasses in broad-leaved crops (Figure 3.2), such as soya beans, potatoes and carrots. As the molecule contains a chiral centre (the carbon bearing the H, CH_3, aromatic and ester groups), the compound will exist as a pair of enantiomers. A decade or so ago it would have been marketed as a racemic mixture (50/50) of the two enantiomers, but the trend nowadays is to market only the enantiomer that is biologically active as the pesticide. This is more expensive to produce than the racemic mixture. However, the advantages are that less material needs to be applied to the crop, and there are less likely to be any adverse environmental effects. You will recall the same factors discussed with reference to the synthesis of pseudoephedrine in Part 3.

Figure 3.2 Selective application of the herbicide Fusilade™ to a strawberry crop.

Fusilade™ is the *R*-enantiomer, and the latter stages in its manufacture involve a nucleophilic substitution reaction with 2-chloropropanoic acid (**3.1**):

(3.1)

3.1

The required *S*-enantiomer of **3.1** (shown above) is obtained in one of two ways. One involves chlorinating the appropriate enantiomer of lactic acid (**3.2**). Alternatively, the racemic 2-chloropropanoic acid is fed to particular soil microbes, which conveniently destroy the unwanted enantiomer. The substitution of *S*-2-chloropropanoic acid involves an S_N2 reaction, in which inversion of configuration gives the desired Fusilade product. In the final step, this product is esterified with butan-1-ol to give Fusilade.

3.2

Fusilade also illustrates another general point about modern pesticides — their high activity or potency. It is active at a level of only 250 g per hectare. Since a hectare is 100 m × 100 m, this is equivalent to spreading this small amount of chemical over two football pitches!

Interestingly, some related compounds — the phenoxyalkane carboxylic acids and esters (**3.3**) — containing a similar grouping (highlighted in red) to that in Fusilade™, have quite the opposite selectivity. They kill broad-leaved weeds — for example, charlock — growing in grass and cereal crops. Indeed, you may have unknowingly used them yourself, since most weed-and-feed products for treating domestic lawns contain a compound of this type for killing daisies and dandelions.

3.3

These herbicides are marketed as the racemate, and have been used extensively by cereal farmers for over 60 years without any major problems. Virtually all land used for growing cereal crops in the UK is treated with a phenoxyalkane carboxylate herbicide.

As already indicated, these herbicides are very selective. This is a major requirement for any pesticide nowadays, since almost all of the well-publicized difficulties associated with a few problem pesticides were a consequence of their lack of selectivity, leading to unexpected side-effects. The classic example of these is the insecticide DDT (Figure 3.3).

Figure 3.3 Children in a residential area of San Angelo, Texas watch health employees spray DDT to combat an increase in polio cases in 1949.

3.1.2 Penicillins

Penicillins were one of the first antibiotics, and continue to be a major weapon in the fight against bacterial infections. They are prepared by fermentation in enormous stainless steel fermenters with capacities up to 100 000 litres (Figure 3.4). The 'synthesis' is carried out by the micro-organism *Penicillium chrysogenum*, which was originally isolated from a mould growing on a melon in a Mexican market! The fermentation broth produces the penicillin in a concentration of about 25 g litre^{-1}.

By adjusting the so-called 'side-chain acid' added to the broth, the R^1 group in the penicillin (see Figure 3.1) can be varied; for example, adding $C_6H_5OCH_2COOH$ to the broth gives penicillin V, in which $R^1 = C_6H_5OCH_2-$.

A key breakthrough in the synthesis of penicillins came in the early 1960s, when research chemists at Beechams decided to investigate a puzzle. In biological and chemical assays for penicillin the results were always different, the chemical result always being higher. The Beechams chemists speculated that this was due to some material being present, which, although having a penicillin structure, was not active as an antibiotic. Furthermore, knowing the penicillin structure, they suggested that the compound was 6-aminopenicillanic acid (6-APA, **3.4**), and that it was formed from the amino acids cysteine and valine (Reaction 3.2). Within days, they confirmed this by carrying out a fermentation in the absence of any side-chain acid.

Figure 3.4
Modern fermentation vessel used for the production of penicillins.

$$H_2N-CH-CH_2-SH$$

cysteine

$$+ \; H_2N-CH_2-COOH \longrightarrow$$

valine

(3.2)

3.4

They could never have imagined the enormous commercial consequences of having solved this puzzle, which opened up the field of semi-synthetic penicillins. Starting from 6-APA, by reaction with an appropriate acyl chloride, RCOCl, several thousand semi-synthetic penicillins were synthesised and tested. This enormously expanded the range and properties of biologically active penicillins (Figure 3.5).

Interestingly, 6-APA is not manufactured by the route shown above, but by first making penicillin G (**3.5**) and then hydrolysing off the $C_6H_5CH_2CO$ group:

$$C_6H_5CH_2CONH \qquad \xrightarrow{\;H_2O\;} \qquad H_2N$$

(3.3)

3.5   **3.4**

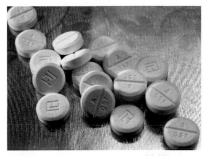

Figure 3.5
Penicillin tablets.

3.1.3 Procion dyes

Procion dyes (Figure 3.6) were the first fibre-reactive dyes. They impart great colour fastness; that is, they are not likely to be washed out. This occurs because the dye reacts with the cotton fibre, which is mainly the biopolymer cellulose, **3.6**:

3.6

Cellulose contains many OH groups, and these can react at each of the two carbon atoms attached to chlorines in the procion dye (Figure 3.1) to give the covalent bond attachment shown in **3.7**.

3.7

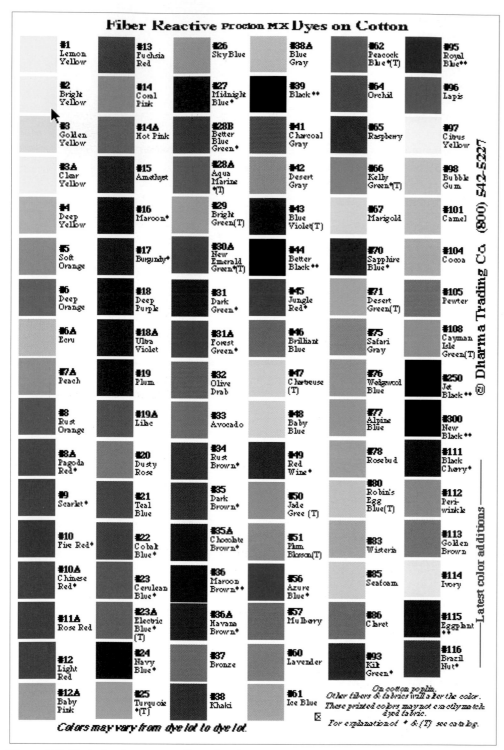

Figure 3.6 Typical colour range of procion dyes.

3.1.4 Kevlar

Speciality polymers have been developed particularly for the aerospace industries. Kevlar[TM]* (Figure 3.7) is a synthetic polyamide. The nylons, such as nylon-6,6 (**2.1**), have CH_2 groups in between the amide groups. However, Kevlar has aromatic

* The structure and applications of Kevlar are discussed in the Case Study in *The Third Dimension*[3].

rings between the amide groups, as shown in Figure 3.1. Hence, it is an example of an aromatic polyamide — a series known as *aramids*. Kevlar is composed of rigid benzene rings joined by amide groups in a linear chain. Its polymer chains adopt an ordered structure with hydrogen-bonding between the amide groups in the adjacent polymer chains keeping the molecules securely aligned — rather like logs in a river. This results in the Kevlar fibres being stiff and strong, which renders it suitable for use in bulletproof vests, skis and aircraft structures. Its excellent thermal stability (it is stable to over 400 °C) accounts for its use as an asbestos replacement in brake linings and gaskets. Remarkably, on a mass-for-mass basis, it is five times stronger than steel.

Note that all the compounds discussed in Section 3 have fairly complex structures. In contrast, we saw that petrochemicals have very simple structures, which do not usually have any chiral centres in the molecule, nor the possibility of geometrical isomerism. This makes their manufacture much easier than that of speciality and fine chemicals.

Figure 3.7
A bulletproof vest made from Kevlar.

3.2 Characteristics of the speciality and fine chemicals sector

The products of this sector have a high to very high value per unit of weight. As an extreme example, the anti-ulcer drug Zantac™, Ranitidine, **3.8** (Figure 3.8), was the world's biggest selling drug for several years, at greater than £3 million per tonne — far more precious than gold or platinum. Since each pill will contain only milligrams of Zantac, 1 tonne will produce a vast number of pills! A more average figure for other fine chemicals would be £10 000 to £100 000 per tonne. Even the common pharmaceutical aspirin sells at a price equivalent to £15 000 per tonne.

3.8

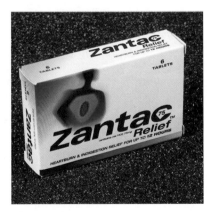

Figure 3.8
Zantac (Ranitidine) medicine.

One reason for the high selling price is the high purity required for these products. Clearly, even small amounts of an impurity in a drug could cause undesirable side-effects. For this reason, quality control is a very significant cost (greater than 10% of total costs) in the production of pharmaceutical products.

The products are produced on a small scale, typically in the range 10–100 tonnes per annum. This usually involves **batch** production; that is, the reactants are loaded into the reactor, the reaction is carried out, and the products are isolated, much in the same way that reactions are carried out in a chemistry laboratory. One consequence of this is that the plants can be 'multi-product'; that is, during the year several different products will be made in the one plant. Due to the small scale of operation, effluent from the process is not usually a significant concern, although it must still be suitably treated and rendered harmless before being discharged.

Research and development (R&D) is a major activity; indeed it has been described as the lifeblood of the pharmaceutical sector, with over 20% of sales *income* being ploughed back into R&D each year. This makes it one of the most research-intensive sectors of manufacturing industry, exceeded only by the electronics and aerospace industries. With problems of organism resistance and the need to reduce side-effects, the search for new drugs and pesticides is ongoing. In many ways, these areas have become victims of their own success. Due to the great efficiency of antibiotics and vaccines, several diseases, which were major life takers at the beginning of the last century, have either been tamed or virtually eradicated; examples are diphtheria, smallpox and tuberculosis (TB). Another striking example has been the use of penicillins to prevent septicaemia. The enormous value of this is apparent when one realizes that septicaemia caused by wound infection was responsible for more deaths during the First World War than action on the battlefield. Mention could also be made of the billions of lives which the infamous insecticide DDT saved through its use in malaria eradication programmes before its withdrawal because of undesirable environmental effects.

As a result of these successes, people are living much longer.
In consequence, diseases associated with old age, such as Alzheimer's, Parkinson's, senile dementia and high blood pressure, which were formerly insignificant, have now come to the fore and are major targets for the pharmaceutical industry. 'New' diseases such as AIDS, and old but complex diseases like cancer, are also major targets. The problems of resistance are well illustrated by tuberculosis, which was virtually conquered in the 1970s, but now
the number of cases in some countries (such as Russia) is rising rapidly. Older antibiotics are now ineffective due to the build up of resistance, promoting the need for new alternatives. Thus, there will always be new diseases to target, and a need to improve the efficiency of existing products in terms of reducing their side-effects and overcoming bacterial resistance to them. Both these factors ensure that there will be a continuing enormous investment in research and development. Indeed, the very largest of the pharmaceutical companies now each invest over £1 billion per annum in R&D!

3.3 Chemical considerations

As we have seen, chiral drugs and pesticides are being increasingly marketed as a single active enantiomer, whereas in the past the racemate would have been supplied. Lessons learned from the past are one of the reasons for this; a notable example is the thalidomide™ tragedy in the late 1950s*. This drug was administered to pregnant women as an anti-nausea drug, and had the advantage over the alternative barbiturates that it was non-addictive. A tragic side-effect was children being born with birth defects. Subsequent research showed that this was caused by the other (inactive) enantiomer which was present in the racemate.

As pointed out previously, marketing only a single enantiomer means that less compound is needed (reducing costs!), and adverse environmental or side-effects are less likely. Set against this will be higher production costs than for the racemate. This is because either special, more-expensive chiral reagents, which very much favour the synthesis of one enantiomer, will be used, or separation of enantiomers will be necessary, which is not straightforward. This is over and above the normal steps in the synthesis, as has already been illustrated in the discussion on Fusilade.

The chemistry involved in the synthesis of speciality and fine chemicals is typically 'academic' type chemistry, and can therefore involve expensive, sophisticated reagents. For example, alkenes react with osmium tetroxide such that two OH groups are added across a double bond. Importantly, both OH groups are added to the same side of the double bond:

(3.4)

This reaction could be used in the synthesis of a pesticide or pharmaceutical, where their high value can bear the cost of the reagent. In contrast, this would not be possible in the petrochemicals sector, where the products only command several hundred pounds per tonne. Here, oxidations such as this would often use air (free!) over a suitable catalyst. However, the toxicity of osmium tetroxide would also militate against its use on a large scale.

3.4 Location of plants

The relatively small scale of production coupled with the high value per unit weight of the products means that there is little restriction on the geographical location of plants. In addition, transport costs are not a significant item. Without such limitations, pharmaceutical plants can be built in any desired location, such as the Kent coast, just south of the Lake District, in mid-Cheshire and in Hertfordshire. This contrasts with petrochemical plants that are only located at the mouths of river estuaries.

* The thalidomide story is discussed in Part 2 of *The Third Dimension*[3].

WHAT HAPPENS WHEN THE FOSSIL FUELS RUN OUT?

4

The previous discussion has highlighted the importance of fossil carbon sources such as natural gas, crude oil and coal, all of which are not renewable. These fossil fuels are converted into a number of key intermediates that are used to make polymers, paints, etc. The intermediate chemicals can also be converted into speciality chemicals such as pharmaceuticals, pesticides and dyes. But, what happens when the fossil fuel sources run out?

Perhaps it would be sensible to start to use renewable sources as soon as possible before we reach this state. One renewable source of carbon compounds is plants, which produce organic compounds from carbon dioxide and water, using energy from sunlight. Many of these materials are polymeric and are already in everyday use; examples are polyamides in silks and wool, and polyalkenes in natural rubber. A more ubiquitous material is carbohydrates, which are used in plants as energy stores and as structural materials. Carbohydrates are polymers such as sugars, containing many OH functional groups (see Structure **3.6**). These polysaccharides are probably more familiar to you as the polymers found in linen and cotton. Many of these materials are used without any chemical treatment, but it is possible to chemically modify them to give other useful materials. Reaction of cellulose with acetic anhydride converts each OH group in the cellulose into an acetate ester. Cellulose acetate has a range of uses; for many years it was used for safety films, and nowadays it is used in lacquers, textiles, transparent sheeting and cigarette filters. Plants also make oils and fats. Although these are used mainly in the food industry, they are also the source of a number of long-chain carboxylic (fatty) acids, alcohols and their derivatives. The sodium salts of the fatty acids are the main ingredients of soaps.

Although these natural materials are useful alternatives to petrochemicals, they would be more useful if they could be converted into the key intermediates such as ethylene (ethene) or propylene (propene). One strategy is to ferment plant material to give methanol or ethanol. Both of these alcohols have been used as fuel sources. In some countries (such as Brazil), alcohol obtained by fermentation is added to petrol. The fuel for conventional cars in Brazil must contain at least 15% alcohol, and a 96% alcohol product is available for use in specially modified car engines. This has been done to reduce the country's dependence on oil imports. In some states in the USA, methanol is sold in garages alongside petrol (or gasoline). It has the advantage that it can be used in most internal combustion engines without too much modification. It can also be used as an alternative to diesel (Figure 4.1). Ethanol can also be used to make ethylene by elimination of water, which can then be used as described earlier to make petrochemicals.

So, although oil is the current dominant source of organic chemicals, this is likely to change in the future. As reserves of oil and coal are depleted, sources of carbo-hydrates will become more important, especially if they can generate the same set of intermediates as crude oil. Thus, in future, the farmers may be growing crops to feed the petrochemical industry.

Figure 4.1 Bus that runs on methanol in California.

FURTHER READING

1   L. E. Smart (ed.), *Separation, Purification and Identification*, The Open University and the Royal Society of Chemistry (2002).

2   M. Mortimer and P. G. Taylor (eds), *Chemical Kinetics and Mechanism*, The Open University and the Royal Society of Chemistry (2002).

3   L. E. Smart and J. M. F. Gagan (eds), *The Third Dimension*, The Open University and the Royal Society of Chemistry (2002).

ACKNOWLEDGEMENTS

Grateful acknowledgement is made to the following sources for permission to reproduce material in this book:

Figure 1.1: courtesy of Ford Motor Company; *Figure 2.1*: courtesy of Exxon; *Figures 2.3, 2.5 and 2.16*: © BP plc (2002); *Figure 2.9*: BP Barex; *Figure 2.12*: courtesy of Combat Polystyrene Packaging Products; *Figure 2.14*: Adelheid Raqué-Nuttall; *Figure 3.2*: courtesy of Syngenta Crop Protection (UK) Limited; *Figure 3.3*: © Betmann/Corbis; *Figure 3.4*: Ed Young/SPL; *Figure 3.5*: Cordelia Molloy/SPL; *Figure 3.7*: courtesy of E. I. Dupont de Nemours & Company; *Figure 3.8*: Mike Levers/Glaxo SmithKline Beecham; *Figure 4.1*: courtesy of Warren Gretz/NREL.

Every effort has been made to trace all the copyright owners, but if any has been inadvertently overlooked, the publishers will be pleased to make the necessary arrangements at the first opportunity.

INDEX

Note Principal references are given in bold type; picture references are shown in italics.

A

acetone (propanone), production of, 154, 156
acetylcholine, 94, 95
acrylic acid, 152
acrylonitrile, 152
activation of benzene ring, **57**–8, 66, *67*
acylonium ion, 58
addition reactions, 11, 91
    see also syn-addition reactions
adhesives, 26
adrenaline, 96, 97, 99
agonist, 96
    see also noradrenaline
alcohols, manufacture from alkenes, 17
alizarin, 71
alkanes,
    cracking of, 142, 144, 148, 150
    reforming of, 144, 151
alkene hydrobromination, stereochemistry of 106–8
alkenes,
    addition reactions of, 11–27
    hydration reactions of, 17
    hydrogenation reactions of, 28–31
    preparation of *Z*-isomers of, 29, 109
    production of, from crude oil, 144, 146, 150, 152
    reactions of,
            with halogens, 18–27
            with hydrogen halides, 13–18
            with osmium tetroxide, 169
alkylated benzenesulfonates, 152
alkylbenzenes, preparation of, 54–9, 75
alkynes, hydrogenation of, 29, 109, 111
aluminium chloride,
    as Friedel–Crafts catalyst, 54–5, 59, 160
    as halogenation catalyst, 50–1
    in supported catalysts, 162
β-aminoalcohols (2-aminoalcohols), 97
    synthesis of, 122–3
aminobenzene; *see* aniline
Anastas, P. T., 128

angina, 98
aniline (aminobenzene), 68–9
antibiotics, 92, 164–5, 168
antiseptic surgery, 41
aramids, 167
aromatic compounds, 40–1
    production of, from crude oil, 144, 151
    see also electrophilic aromatic substitution reactions
aromaticity, 43–5
aspirin, 92,
asthma therapy, 96
atom economy, 125–6, 126–7
autonomic nervous system, 93–4
Avogadro's constant (Avogadro number), 113
Avogadro's hypothesis, 112
azo dyes, 70, 71, 72
AZT (zidovudine), 92

B

batch production, 168
benzene,
    as product of catalytic reforming, 151
    as starting material in petrochemicals industry, 156–7
    from coal tar, 40, 41
    from crude oil, 151
    reactions of,
            Friedel–Crafts, 53–9
            halogenation, 50–1
            nitration, 48–9
            sulfonation, 51–3
    structure and stability of, 42–4
benzenediazonium salts, 70–2, 75–6
benzenesulfonic acid, preparation of, 51–3, 75
benzoic acid, preparation of, 76
benzonitrile, preparation of, 76
beta-blockers, 98
bisphenol A, 26
Black, Sir James, 98
bromoalcohols; *see* bromohydrins
bromobenzene, preparation of, 51, 75, 76
N-bromoethanamide (*N*-bromoacetamide), 24, 114

bromoethers, preparation of, 23, 24
bromohydrins (bromoalcohols), **23**–5, 105, 106
    preparation of, 114–17
bromonium ions; *see* cyclic bromonium ions
buta-1,3-diene, 151, 152, 155
trans-but-2-ene, reaction of, with bromine of, 20–2

C

Calvert, F. C., 41
captopril, 92
carbohydrates, 165, 170
carbolic acid, 41
    see also phenol
Carré, Edmond, 49
catalysts,
    Friedel–Crafts, 54–5, 58, 59, 160
    halogenation, 50–1
    hydrogenation, 28, 30, 31, 109, 111, 126, 127
    'poisoned', 109, 111
    see also 'Envirocats™'
catalytic reagents, 128
chemical industry, 139–43
    waste treatment and disposal in, 125, 162
    see also petrochemicals industry; speciality and fine chemicals
chiral compounds, 99, 118
chlorobenzene,
    nitration of, 61
    preparation of, 50, 75, 76
cisplatin, 92
Clemmensen reduction, 59
coal, 39, 40
    feedstock for chemical industry, 144–5
coal gas, 39, *40*, 145
coal tar, 39–40, 145
    dyes from, 68–71
    use in surgery, 41
combinatorial chemistry, 93, 123
coupling (diazonium salts), **72**–3, 76
cracking (crude oil), 142, **144**, 148, 150
Crafts, James Mason, 54
crude oil, 142, 144, 145
    composition of, 147, *148*
    cracking of, 142, **144**, 148, 150
    feedstock for chemical industry, 144–6
    fractionation of, 147–8, *149*
    reforming of, 144, 151
crude yield of reaction, **113**

cumene (2-phenylpropane),
    manufacture of, 55–7, 154, 156
    preparation of, 55
curly arrows, how to use, 12
cyclic bromonium ions, 20–5, 106, 107

D

DDT, 164, 168
deactivation of benzene ring, **58**, 66–7
delocalized π electron systems, 43–4, 46
desulfonation, 52–3
detergents, synthetic, 152
diastereoisomers, separation of, 119–20
diazonium salts, 68–74, 75, 76
    coupling reactions of, 72–3, 76
    substitution reactions of, 73–4, 76
diazotization, 71–2
meso-2,3-dibromobutane, preparation of, 21–2
digitalin, 92
1,2-dihalides, preparation of, 18, 20–2
para-dimethylaminophenylazobenzene, preparation of, 72
directing effects, 62–7
double bonds,
    migration of, 18
    see also π-bonds
drugs, 71, 91, 167–9
    binding of, to receptors, 94
    design and synthesis of, 91–3
    resistance of bacteria, to, 92, 168
dyes, 68–71
    procion, 165, *166*

E

Ehrlich, Paul, 71
electrophiles, 11
electrophilic aromatic substitution reactions, 43, **45**–7, 91
    effects of substituents in, 61–7
    Friedel–Crafts reactions, 53–9, 160
    halogenation, 50–1
    nitration, 48–9, 81
    sulfonation, 51–3
elimination reactions, 91
enantiomers,
    manufacture of single, 163, 169
    separation (resolution) of, 118, 119
energy use in petrochemicals industry, 147
'Envirocats™', 160
ephedrine, 96, 97–8, 104
epichlorohydrin, 26
epoxides; *see* oxiranes

epoxy resins, 26
ethanol, plant-derived, 170
ethene (ethylene), 142–3, 151, 152
    electron distribution in, *12*
    from ethanol, 170
    as starting material in petrochemicals industry, 153

F

fats and oils, 30, 170
    hydrogenation of, 30–1
    iodine number of, 19
    see also lipid membranes
fatty acids, 19, 30
fermentation antibiotic production by, 164–5
fine chemicals; *see* speciality and fine chemicals
fluorobenzene, preparation of, 76
formula mass, 125–6
fossil fuels, replacement of, 170
fractional crystallization, 119
Friedel, Charles, 54
Friedel–Crafts acylations, 53, **58**–9
Friedel–Crafts alkylations, 53, **54**–8
Friedel–Crafts reactions, 53, 160
    see also Friedel–Crafts acylations; Friedel–Crafts alkylations
Fusilade™, 162–3

G

gas lighting, 39
Gay-Lussac's law, 112
green chemistry movement, 125–9, 160
Griess, Peter, 72
Gustavson, G, 56

H

halogenation reactions, **50**–1
herbicides, selective, 162–3
homochiral substances, **119**
hydrogen halides, reactions of, with alkenes, 13–18
hydrogenation reactions, 28–31, 109, 111–13
para-hydroxyphenylazobenzene, preparation of, 72–3, 76
hypobromous acid, 24–5, 105, 107, 115–16, 124–5

I

ibuprofen, manufacture of, 126–8
intermediates for chemical industry, 151–2
intermolecular forces, 95
iodine number, 19
iodobenzene, preparation of, 76

K

Kekulé, Friedrich August, 42–3
Kevlar™, 166–7

L

LeBeuf, Ferdinand, 41
Lemaire, Jules, 41
Lewis acids, 50–1, 55
Lindlar's catalyst, 29, 109, 111
linoleic acid, 19
lipid membranes, 97
Lister, Joseph, 41

M

margarine, 30–1
Markovnikov, V. V., 16
Markovnikov's rule, 16, 17
    see also mechanistic Markovnikov rule
mauveine, 68–9, 70
mechanistic Markovnikov rule, 16, 23, 24, 106
mescaline, 100
meta **isomers, 78**–9
methanol, plant-derived, 170, *171*
2-methoxy-2-methylpropane; *see* MTBE
methoxybenzene, nitration of, 63
methyl tertiary butyl ether; *see* MTBE
methylaniline isomers, 69
methylbenzene (toluene), 157
    as starting material for petrochemical industry, 157, 158
    chlorination of, 62
    para substitution of, 80, 81
1-methyl-2-phenyloxiranes,
    preparation of, 105–6, 117
    reactions of, with methylamine, 104–5, 118
2-methylpropan-2-ol, manufacture of, 17
2-methylpropene,
    reactions of,
        hydration, 17
        with hydrogen halides, 14–15
molar mass, 112
molar mass calculator, 112
molar percentage yield calculator, 114
molar scale reactions, **111**
moles,
    of gas, 112
    use in calculations of, 112
MTBE (methyl tertiary butyl ether, 2-methoxy-2-methylpropane), 160

CD-ROM INFORMATION

Computer specification

The CD-ROMs are designed for use on a PC running Windows 95, 98, ME or 2000. We recommend the following as the minimum hardware specification:

processor	Pentium 400 MHz or compatible
memory (RAM)	32 MB
hard disk free space	100 MB
video resolution	800 × 600 pixels at High Colour (16 bit)
CD-ROM speed	8 × CD-ROM
sound card and speakers	Windows compatible

Computers with higher specification components will provide a smoother presentation of the multimedia materials.

Installing the CD-ROMs

Software must be installed onto your computer before you can access the applications. Please run INSTALL.EXE from either of the CD-ROMs.

This program may direct you to install other, third party, software applications. You will find the installation programs for these applications in the INSTALL folder on the CD-ROM. To access all the software on the CD-ROM, you must install, Isis/Draw, WebLab ViewerLite and Acrobat Reader.

Running the applications on the CD-ROM

You can access *Alkenes and Aromatics* CD-ROM applications through a CD-ROM Guide (Figure C.1), which is created as part of the installation process. You may open this from the **Start** menu, by selecting **Programs** followed by **The Molecular World**. The CD-ROM Guide has the same title as this book.

The *Data Book* is accessed directly from the **Start | Programs | The Molecular World** menu (Figure C.2), and is supplied as an Adobe Acrobat™ document.

Problem solving

The contents of this CD-ROM have been through many quality control checks at the Open University, and we do not anticipate that you will encounter difficulties in installing and running the software. However, a website will be maintained at

    http://the-molecular-world.open.ac.uk

which records solutions to any faults that are reported to us.

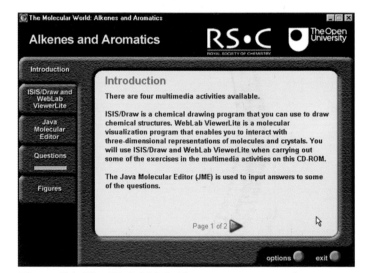

Figure C.1 The CD-ROM Guide.

Figure C.2 Accessing the *Data Book* and CD-ROM Guide.